Fungal Biology

Series Editors

Vijai Kumar Gupta
Biorefining and Advanced Materials Research Center
Scotland's Rural College (SRUC), SRUC Barony Campus, Parkgate
Dumfries, Scotland, United Kingdom

Maria G. Tuohy
School of Natural Sciences
National University of Ireland Galway
Galway, Ireland

About the Series

Fungal biology has an integral role to play in the development of the biotechnology and biomedical sectors. It has become a subject of increasing importance as new fungi and their associated biomolecules are identified. The interaction between fungi and their environment is central to many natural processes that occur in the biosphere. The hosts and habitats of these eukaryotic microorganisms are very diverse; fungi are present in every ecosystem on Earth. The fungal kingdom is equally diverse, consisting of seven different known phyla. Yet detailed knowledge is limited to relatively few species. The relationship between fungi and humans has been characterized by the juxtaposed viewpoints of fungi as infectious agents of much dread and their exploitation as highly versatile systems for a range of economically important biotechnological applications. Understanding the biology of different fungi in diverse ecosystems as well as their interactions with living and non-living is essential to underpin effective and innovative technological developments. This series will provide a detailed compendium of methods and information used to investigate different aspects of mycology, including fungal biology and biochemistry, genetics, phylogenetics, genomics, proteomics, molecular enzymology, and biotechnological applications in a manner that reflects the many recent developments of relevance to researchers and scientists investigating the Kingdom Fungi. Rapid screening techniques based on screening specific regions in the DNA of fungi have been used in species comparison and identification, and are now being extended across fungal phyla. The majorities of fungi are multicellular eukaryotic systems and therefore may be excellent model systems by which to answer fundamental biological questions. A greater understanding of the cell biology of these versatile eukaryotes will underpin efforts to engineer certain fungal species to provide novel cell factories for production of proteins for pharmaceutical applications. Renewed interest in all aspects of the biology and biotechnology of fungi may also enable the development of "one pot" microbial cell factories to meet consumer energy needs in the 21st century. To realize this potential and to truly understand the diversity and biology of these eukaryotes, continued development of scientific tools and techniques is essential. As a professional reference, this series will be very helpful to all people who work with fungi and should be useful both to academic institutions and research teams, as well as to teachers, and graduate and postgraduate students with its information on the continuous developments in fungal biology with the publication of each volume.

More information about this series at http://www.springer.com/series/11224

S. Chandra Nayaka • Rajashekara Hosahatti
Ganesan Prakash • C. Tara Satyavathi
Rajan Sharma
Editors

Blast Disease of Cereal Crops

Evolution and Adaptation in Context of Climate Change

Springer

Editors
S. Chandra Nayaka
Pearl Millet Pathology Laboratory
DOS in Biotechnology
University of Mysore
Manasagangotri, Mysore, India

Ganesan Prakash
Division of Plant Pathology
ICAR-Indian Agricultural Research Institute
New Delhi, India

Rajan Sharma
International Crops Research Institute
for the Semi-Arid Tropics (ICRISAT),
Patancheru
Hyderabad, Telangana, India

Rajashekara Hosahatti
ICAR-VPKAS, Crop Protection Division
Almora, Uttarakhand, India

C. Tara Satyavathi
ICAR-AICRP on Pearl Millet
Jodhpur, Rajasthan, India

ISSN 2198-7777 ISSN 2198-7785 (electronic)
Fungal Biology
ISBN 978-3-030-60584-1 ISBN 978-3-030-60585-8 (eBook)
https://doi.org/10.1007/978-3-030-60585-8

This Springer imprint is published by the registered company Springer Nature Switzerland AG
The registered company address is: Gewerbestrasse 11, 6330 Cham, Switzerland

Preface

The edited book is dedicated to pathology research workers which mainly focuses on ***Blast Disease of Cereal Crops: Evolution and Adaptation in Context of Climate Change*** focuses on blast disease of different cereal crops and its management through integrated disease management (IDM) and genomic approaches. Blast disease caused by ascomycete's hemibiotrophic pathogen is an important foliar disease infecting a majority of cereal crops like rice, finger millet, pearl millet, foxtail millet, wheat, and other cereal crops, and causes a huge economic impact globally. The pathogen is responsible for causing many devastating epidemics in many crops over the years and shifting to new hosts is most common nowadays. *Magnaporthe* spp. is the most prominent cause of blast disease on a broad host range of grasses, as well as other species of poaceae family. To date, 137 members of poaceae hosting this fungus have been described in fungal databases. In the past two decades, there have been significant developments in genomics and proteomics approaches, and there have been substantial and rapid progress in cloning and mapping of R genes for blast resistance, and comparative genomics analysis for resolving species delineation of *Magnaporthe* spp. infecting both cereals and grass species. Blast disease resistance follows a typical gene-for-gene hypothesis. Identification of new *Avr* genes and effector molecules from *Magnaporthe* spp. can be useful to understand the molecular mechanism involved in the fast evolution of different races of this fungus. It may help to reduce the occurrence of quick breakdown of blast resistance by the identification of potential R genes for effective deployment.

This book provides information on all blast diseases infecting different cereal crops in different aspects, like how the fast evolution of pathogens is due to high variability in a given time and space leading to adapting to new hosts and causing epidemics in a short amount of time. It covers different aspects like symptomatology, casual organism, historical perspective, pathogen evolution, occurrence, epidemiology, microconidia as its role in inciting plant disease, use of chemical fungicides, host range shift, cross-infectivity, isolation of pathogen, blast disease infecting cereal crops in context of climate change, genomics, and genome editing approaches which potentially helps in understanding the nature of R genes and applying the emerging concept and technologies which can make real impact in sustainable management of blast disease in different cereal crops.

We are grateful to all the contributors for their valuable book chapters and kind cooperation. We are also thankful to the entire team of workers at Springer for their guidance and help in publishing this book. This text should serve as a reference guide for scientists, teachers, students, scholars, administrators, and policy makers dealing with blast disease of crops and their disease management.

Manasagangotri, Mysore, India	S. Chandra Nayaka
Alomora, Uttarakhand, India	Rajashekara Hosahatti
New Delhi, India	Ganesan Prakash
Jodhpur, Rajasthan, India	C. Tara Satyavathi
Hyderabad, Telangana, India	Rajan Sharma

The original version of this book was revised. The co-editor name Ganesan has been updated. The correction to this chapter can be found at https://doi.org/10.1007/978-3-030-60585-8_15

Contents

Contributors

C. Amoghavarsha Department of Plant Pathology, University of Agricultural and Horticultural Sciences, Shivamogga, Karnataka, India

Y. Rupert Anand College of Agricultural Engineering and Post-Harvest Technology, Ranipool, Gangtok, Sikkim, India

B. M. Anil Kumar Pearl Millet Pathology Laboratory, DOS in Biotechnology, ICAR-AICRP, Mysore Centre, University of Mysore, Mysore, India

N. Anuradha Agricultural Research Station, ANGRAU, Gajularega, Vizianagaram, Andhra Pradesh, India

K. Arora ICAR-National Institute for Plant Biotechnology, New Delhi, India

P. K. Bhowmick ICAR-Indian Agricultural Research Institute, New Delhi, India

S. Bhuvaneshwari ICAR Research Complex for NEH Region, Manipur Centre, Lamphelpat, Imphal, India

S. Chandra Nayaka Pearl Millet Pathology Laboratory, DOS in Biotechnology, University of Mysore, Manasagangotri, Mysore, India

E. Chidanandappa Rice Pathology Laboratory, All India Coordinated Rice Improvement Programme, University of Agricultural Sciences, Raichur, Karnataka, India

I. K. Das ICAR-Indian Institute of Millets Research, Rajendranagar, Hyderabad, Telangana, India

B. N. Devanna ICAR-National Rice Research Institute, Cuttack, Odisha, India

Abhishek Kumar Dubey Division of Crop Research, ICAR-Research Complex for Eastern Region, ICAR Parisar, Patna, Bihar, India

Saurabh Kumar Dubey Division of Plant Pathology, ICAR-Indian Agricultural Research Institute, New Delhi, India

R. K. Ellur ICAR-Indian Agricultural Research Institute, New Delhi, India

K. E. Georgia Agricultural Research Station, ANGRAU, Gajularega, Vizianagaram, Andhra Pradesh, India

S. Gopala Krishnan ICAR-Indian Agricultural Research Institute, New Delhi, India

B. Haritha ICAR-Indian Agricultural Research Institute, New Delhi, India

Rajashekara Hosahatti ICAR-Vivekananda Institute of Hill Agriculture, Almora, India

P. Jain ICAR-National Institute for Plant Biotechnology, New Delhi, India

B. Jeevan ICAR-Vivekananda Parvatiya Krishi Anusandhan Sansthan, Almora, Uttarakhand, India

Lakshmi Kant ICAR-Vivekananda Parvatiya Krishi Anusandhan Sansthan, Almora, Uttarakhand, India

J. L. Katara ICAR-National Rice Research Institute, Cuttack, Odisha, India

A. Khanna ICAR-Indian Agricultural Research Institute, New Delhi, India

A. Kumar ICAR-National Rice Research Institute, Cuttack, Odisha, India

Aundy Kumar Division of Plant Pathology, ICAR-Indian Agricultural Research Institute, New Delhi, India

Sudhir Kumar ICAR-Indian Institute of Agricultural Biotechnology (IIAB), Namkum, Ranchi, Jharkhand, India

B. Malikarjuna Project Coordinating Unit, ICAR-AICRP on Small Millets, UAS, GKVK, Bengaluru, Karnataka, India

Sudarshan Maurya ICAR-Indian Institute of Vegetable Research, Varanasi, India

Ram Prasnna Meena ICAR-Directorate of Medicinal and Aromatic Plants Research, Boriavi, Anand, Gujarat, India

K. K. Mishra ICAR-Vivekananda Parvatiya Krishi Anusandhan Sansthan, Almora, Uttarakhand, India

Sajad Un Nabi ICAR-ICAR-Central Institute of Temperate Horticulture, Srinagar, Union Territory of Jammu and Kashmir, India

M. Nagarajan ICAR-Indian Agricultural Research Institute, New Delhi, India

R. Narasimhulu Acharya NG Ranga Agricultural University Agricultural Research Station, Ananthapuram, Andhra Pradesh, India

K. B. Palanna Project Coordinating Unit, ICAR-AICRP on Small Millets, UAS, GKVK, Bengaluru, Karnataka, India

C. Parameswaran ICAR-National Rice Research Institute, Cuttack, Odisha, India

Asharani Patel Division of Plant Pathology, ICAR-Indian Agricultural Research Institute, New Delhi, India

T. S. S. K. Patro Agricultural Research Station, ANGRAU, Gajularega, Vizianagaram, Andhra Pradesh, India

Ganesan Prakash Division of Plant Pathology, ICAR-Indian Agricultural Research Institute, New Delhi, India

Ish Prakash Division of Plant Pathology, ICAR-Indian Agricultural Research Institute, New Delhi, India

D. Pramesh Rice Pathology Laboratory, All India Coordinated Rice Improvement Programme, University of Agricultural Sciences, Raichur, Karnataka, India

M. K. Prasanna Kumar Department of Plant Pathology, University of Agricultural Sciences, GKVK, Bengaluru, Karnataka, India

B. Praveen Project Coordinating Unit, ICAR-AICRP on Small millets, UAS, GKVK, Bengaluru, Karnataka, India

Boda Praveen Project Coordinating Unit, ICAR-AICRP on Small Millets, UAS, GKVK, Bengaluru, Karnataka, India

S. Raghu ICAR-National Rice Research Institute, Cuttack, Odisha, India

A. Raghunandana Rice Pathology Laboratory, All India Coordinated Rice Improvement Programme, University of Agricultural Sciences, Raichur, Karnataka, India

Chandramani Raj ICAR Research Complex for NEH Region, Sikkim Center, Tadong, Gangtok, India

Rajesha G ICAR-Indian Institute of Millets Research, Rajendranagar, Hyderabad, Telangana, India

Gutha Venkata Ramesh Project Coordinating Unit, ICAR-AICRP on Small Millets, UAS, GKVK, Bengaluru, Karnataka, India

H. R. Raveendra ICAR-AICRP on Small Millets, Zonal Agricultural Research Station, UAS (B), V.C. Farm, Mandya, Karnataka, India

A. Saddamhusen Department of Plant Pathology, University of Agricultural Sciences, GKVK, Bengaluru, Karnataka, India

Kuleshwar Prasad Sahu Division of Plant Pathology, ICAR-Indian Agricultural Research Institute, New Delhi, India

Sakthivel Krishnan ICAR-Indian Institute of Oilseed Research, Rajendranagar, Hyderabad, Telangana, India

S. Samantaray ICAR-National Rice Research Institute, Cuttack, Odisha, India

C. Tara Satyavathi ICAR-AICRP on Pearl Millet, Jodhpur, Rajasthan, India

H. Sharanabasav Rice Pathology Laboratory, All India Coordinated Rice Improvement Programme, University of Agricultural Sciences, Raichur, Karnataka, India

Devender Sharma ICAR-Vivekananda Parvatiya Krishi Anusandhan Sansthan, Almora, Uttarakhand, India

Parul Sharma Department of Crop Improvement, College of Agriculture, CSK Himachal Pradesh Krishi Vishvavidyalaya, Palampur, Himachal Pradesh, India

Rajan Sharma International Crops Research Institute for the Semi-Arid Tropics (ICRISAT), Patancheru, Hyderabad, Telangana, India

Susheel Kumar Sharma ICAR Research Complex for NEH Region, Manipur Centre, Lamphelpat, Imphal, India

A. K. Singh ICAR-Indian Agricultural Research Institute, New Delhi, India

Ashish Kumar Singh ICAR-Vivekananda Parvatiya Krishi Anusandhan Sansthan, Almora, Uttarakhand, India

J. Singh ICAR-National Institute for Plant Biotechnology, New Delhi, India

P. K. Singh National Agri-Biotechnology Institute, Mohali, Punjab, India

Rashmi Singh Smt. Indira Gandhi Post Graduate Degree College, Lalganj, Mirzapur, Uttar Pradesh, India

Shweta Singh ICAR Research Complex for NEH Region, Sikkim Center, Tadong, Gangtok, India

Thokchom Surjit Singh ICAR Research Complex for NEH Region, Manipur Centre, Lamphelpat, Imphal, India

A. R. N. S. Subbanna ICAR-Vivekananda Parvatiya Krishi Anusandhan Sansthan, Almora, Uttarakhand, India

Vilas A. Tonapi ICAR-Indian Institute of Millets Research, Rajendranagar, Hyderabad, Telangana, India

R. L. Verma ICAR-National Rice Research Institute, Cuttack, Odisha, India

K. K. Vinod ICAR-Indian Agricultural Research Institute, New Delhi, India

Manoj Kumar Yadav Crop Protection Division, ICAR-National Rice Research Institute, Cuttack, Odisha, India

Abbreviations

ACT-PCR	Annealing at critical temperature—polymerase chain reaction
AFLP	Amplified fragment length polymorphism
ALS	Aceto lactate synthase
ATPs	Adenosine triphosphates
AVR	Avirulence
BC	Before christ
BTH	Acibenzolar-*s*-methyl
CAPS	Cleavage amplified polymorphic sequence
Cas	CRISPR-associated system
CC-NBS-LRR	Coiled coil nucleotide-binding site leucine-rich repeats
CDSs	Coding DNA sequences
CO2	Carbon dioxide
DMIs	DeMethylation inhibitors
DNA	Deoxyribo nucleic acid
DR	Defense regulators
DSB	Double-stranded break
EBDCs	Ethylene bis-dithiocarbamates
EBE	Effector binding element
EDDP	Edifenphos
EMS	Ethyl methyl sulfonate
ERF	Ethylene-responsive factor
ET	Ethylene
ETI	Effector triggered immunity
F_1	First filial generation
F_2	Second filial generation
FRAC	Fungicide resistance action committee
HCN	Hydrogen cyanide
HDR	Homology-directed repair
HE	Homing endonucleases
HR	Hypersensitive response
HRMA	High resolution melting analysis

IBP	Iprobenphos
IH	Invasive hyphae
InDel	Insertion deletion
ISR	Induced systemic resistance
JA	Jasmonic acid
LRR	Leucine-rich repeats
MABB	Marker-assisted backcross breeding
MAPK	Mitogen-activated protein kinase
MAS	Marker-assisted selection
MB	Map based
MB	Mega bases
MBIs	Melanin biosynthesis inhibitors
mha	Million hectare
mlo	Mildew locus
MoL	*Magnaporthe oryzae* Lolium pathotype
MoO	*Magnaporthe oryzae* Oryza pathotype
MoT	*Magnaporthe oryzae* Triticum pathotype
mt	Million tons
mtDND	Mitochondrial DNA
NBS-LRR	Nucleotide binding site leucine-rich repeats
NHEJ	Non-homologous end joining
NILs	Near-isogenic lines
NRPS	Non-ribosomal peptide synthases
PAGE	Polyacrylamide agarose gel electrophoresis
PAMP	Pathogen-associated molecular patterns
PBZ	Probenazole
PCR	Polymerase chain reaction
PDA	Potato dextrose agar
PGPR	Plant growth-promoting rizobacteria
PKS	Polyketide synthases
ppm	Parts per million
PPO	Poly phenol oxidase
PRR	Plant pattern recognition receptors
PTI	PAMP-triggered immunity
PTI	Pathogen-triggered immunity
PTL	Phosphorothiolate
QTL	Quantitative trait loci
R genes	Resistance genes
RAPD	Random amplified polymorphic DNA
rDNA	Ribosomal DNA
RFLP	Restriction fragment length polymorphism
RNA	Ribo nucleic acid
RNAi	RNA interference
ROS	Reactive oxygen species
RPG	Recurrent parent genome

SA	Salicylic acid
SAR	Systemic acquired resistance
SBIs	Sterol biosynthesis inhibitors
SC	Soluble concentrate
SCAR	Sequence characterized amplified region
SH groups	Sulfhydryl groups
SNP	Single nucleotide polymorphism
SSR	Simple sequence repeats
STS	Sequence tagged site
TALE	Transcription activator like effector proteins
TALEN	Transcription activator like effector nucleases
TIR	Toll-interleukin-1 receptor
WG	Wettable granules
WP	Wettable powder
ZFN	Zinc finger endo-nucleases

Chapter 1
Blast Disease: Historical Importance, Distribution, and Host Infectivity Across Cereal Crops

Rajashekara Hosahatti, B. Jeevan, K. K. Mishra, A. R. N. S. Subbanna, and Lakshmi Kant

Contents

1.1 Rice Blast Symptoms

The blast pathogen infects all the growth stages and parts of rice plant including leaf blades, nodes, and neck region (Fig. 1.1); in severe cases, the infection can be seen on leaf sheaths, rachis, joints of the culm, and even on glume. Blast disease can be described in three types depending on symptoms like leaf blast, panicle/neck blast, and node blast.

R. Hosahatti (✉) · B. Jeevan · K. K. Mishra · A. R. N. S. Subbanna · L. Kant
ICAR-Vivekananda Parvatiya Krishi Anusandhan Sansthan, Almora, Uttarakhand, India

S. C. Nayaka et al. (eds.), *Blast Disease of Cereal Crops*, Fungal Biology,
https://doi.org/10.1007/978-3-030-60585-8_1

(A) (B) (C)

Fig. 1.1 Symptoms of rice blast during different growth stages: (**a**) leaf blast, (**b**) panicle blast, (**c**) node blast

1.1.1 Leaf Blast

The initial infections can be seen in the form of minute and brown-colored lesions, specks, or spots which elongate over time to become spindle-shaped pointed lesions at both ends which measure several centimeters long and about 0.5–1.0 cm wide. The center region of spot is greenish gray often covered by a brownish margin. The lesion characteristics, viz., size, shape, and color, will vary with different climatic conditions and also depending on varietal response. With prevailing favorable conditions, the disease progresses on a susceptible cultivar which appears as larger and broader lesions which are more in number and finally coalesce, leading to complete drying of the entire leaf (Padmanabhan 1974; Manibhushanrao 1994).

1.1.2 Panicle Blast

The panicle blast starts appearing at the booting stage of the crop where the stem portion just below the ear becomes brown or black which is a characteristic symptom, the neck blast. The infection is mostly confined to the neck region or sometimes the individual branches of the panicle get infected and turn brown to black. The infected panicles and whole inflorescence often break and fall off at the rotten neck. The grains on the panicle of the infected neck are generally chaffy and look whitish by far distance.

1.1.3 Node Blast

The infection starts at the juncture of two nodes, mostly infected in lower nodes of the rice plants, and turns black at the point of infection. The infected rachis and glumes contain brown to black spots particularly where the disease occurs on branches. Finally, the infection leads to breakdown of the entire tillers of plant at the juncture of node portion resulting in yield loss of affected crop. The incidence of node blast was observed in moderate severity during recent years in North-Western Himalayan region.

1.2 Finger Millet Blast Symptoms

Similar to rice blast, the disease symptoms on finger millet can be seen at all growth stages starting from seedling to grain maturity; depending on the crop growth stage and prevailing environmental conditions the disease can be classified into leaf blast, panicle/neck blast, and finger blast. The symptoms include formation of typical elliptical or diamond-shaped lesions on leaves with gray centers which are water soaked with a chlorotic halo surrounding the lesions. The blast lesions enlarge and coalesce and give burnt appearance depending on congenial conditions. The neck blast symptoms appear as elongated black color lesion usually 1 or 2 in. below the ear and in severe infection toppling of ear head can be observed. Finger blast symptom initially appears as brown spot at the tip and as the disease progresses it proceeds toward the base of the finger. Neck infection is the most destructive stage of the disease that causes major loss in grain yield by decreasing the number and individual grain weight and can also cause spikelet sterility (Fig. 1.2).

1.3 Pearl Millet Blast Symptoms

The initial blast symptoms appear as minute specks or lesions that broaden and turn necrotic, thereby causing widespread chlorosis and complete drying of young leaves (Fig. 1.3). The symptoms are usually referred to as gray leaf spot disease. Initially, the lesions start at the leaf tips or leaf margin or both and extend down along the outer edges. Young lesions are pale green to grayish green in appearance and at later stages they turn into yellow to gray along with maturity of plants. During humid weather conditions and with high plant density the disease becomes severe.

(A) (B) (C)

Fig. 1.2 Symptoms of finger millet blast during different growth stages: (**a**) leaf blast, (**b**) neck blast, (**c**) finger blast

Fig. 1.3 Symptoms of pearl millet blast: (**a**) initial lesions, (**b**) complete drying, (**c**) Setaria leaf blast

1.4 Foxtail Millet Blast Symptoms

The disease is characterized by the presence of diamond- or eye-shaped lesions with gray centers bordered by yellow halo. Over time, these spots enlarge and coalesce to give blasted appearance. Although the symptom appears similar to that of finger millet blast (see Fig. 1.3), the neck and finger infections are almost missing.

1.5 Wheat Blast Symptoms

The symptoms occur on all aboveground parts of the plant. Spindle-shaped lesions, which are water soaked and with gray-green color having dark brown to reddish-brown margin, often have yellow halos, and appear on leaves. These lesions have gray centers during sporulation and white to tan centers after sporulation. As the disease advances these lesions coalesce resulting in complete death of the infected tissue (Rios et al. 2013). The most deleterious stage of disease is seen, if infection during flowering or early grain formation leads to bleaching of the spike. Depending on the susceptibility levels of cultivar, and timing and point of infection, the disease can prevent seed setting or can induce spike sterility. Infections on the rachis or peduncle can also kill the upper parts of the spike which causes highest yield losses (Goulart et al. 2007). Yield loss is in correlation with the extent of spike damage.

1.6 Historical Perspective

The first record of rice blast disease was in China by Soong ying-shin in 1637 in his book on utilization of natural resources (Manibhushanrao 1994). In Japan, it was first reported by Tsuchiya in 1704 (Goto 1955). The causal organism, *Pyricularia oryzae*, was named by Cavara in Italy (Cavara 1892) and subsequently in Japan (Shirai 1896). In Asia, the disease was first reported more than three centuries ago and distributed throughout the rice-growing ecosystems of continents. The pathogen is a complex species, is heterothallic, and is with continuous development of new races and isolates that are diverse in phenotypic virulence (Tharreau et al. 2009). The wide geographic distribution, rapid race evolution, high yield losses, and additional costs incurred in disease management make this disease a serious concern to rice cultivation with an estimated yield loss of US $55 million every year in South and Southeast Asia. The losses are even higher in East Asia and other more temperate rice-growing regions of the world (Herdt 1991). The disease is estimated to cause production loss to an extent of 70–80% (Ou 1985) when predisposition factors (high mean temperature values, relative humidity higher than 85–89%, presence of dew, and excessive nitrogen fertilizer application) favor epidemic development (Piotti et al. 2005). The understanding on biology of rice blast disease is therefore of particular importance, because it promises development of novel and robust disease control strategies (Skamnioti and Gurr 2009).

In India, the disease was first recorded in Thanjavur (Tanjore) delta of South India by Mc Rae in 1918. The disease is causing damage to rice production (Sundararaman 1927; Thomas 1930) in India with its recurrence in every season. But it attracted the attention only when a devastating epidemic occurred in 1919 in the Tanjore delta of erstwhile Madras state (Padmanabhan 1965). Seven epidemics of blast disease occurred incessantly between 1980 and 1987 in the states of Himachal Pradesh, Andhra Pradesh, Tamil Nadu, and Haryana imparting severe yield losses (Sharma et al. 2012).

Finger millet is an important nutri-cereal food crop of rainfed and marginal lands of arid and semiarid regions. It is mostly grown in East Africa, India, and other Asian countries including Sri Lanka and China (Fakrudin et al. 2004). In recent past, overall production and productivity of the crop have been declining majorly due to several biotic and abiotic constraints. Its production is adversely affected by a number of diseases among the blast caused by *M. grisea* (anamorph-*Pyricularia grisea* (Cooke) Sacc.), which is a major problem in India and Africa causing substantial yield losses. The blast on finger millet is known to occur in India (Mc Rae 1920), Sri Lanka (Park 1932), Nepal (Thompson 1941), Malaya (Burnett 1949), Tanzania (Kuwite and Shao 1992), Somalia (Mohamed 1980), Zambia (Muyanga and Danial 1995), Ethiopia, Kenya, and Uganda (Dunbar 1969; Adipala 1992). In India, the blast disease was first recorded from the Tanjore delta of Tamil Nadu by Mc Rae in 1920 with an estimated yield loss of 50% (Venkatarayan 1946). The average grain yield loss was reported to be around 28–36% (Vishwanath et al. 1986; Nagaraja 2007), and in the endemic areas it could be as high as 80–90% (Vishwanath et al. 1986; Bisht 1987; Rao 1990). In finger millet, the host plant resistance to blast disease is often confirmed at the seedling stage, which is not in correlation with neck and finger blast which are economically more damaging. Therefore, both neck and finger are considered as most important parameters of resistance (Nagaraja 2007) which in finger millet can be identified only by screening under natural field conditions (Nagaraja 2007; Nagaraja et al. 2010; Babu et al. 2013). The pure cultures of *M. grisea* are established from infected leaves, necks, and panicles and tested for their pathogenicity and organ specificity toward the other plant parts of finger millet (Puri and Kumar 2012). The cross-infectivity and host range studies with other cereal hosts and weed hosts were carried out with *M. grisea* infection (Shanmugapackiam and Raguchander 2018).

Pearl millet blast is a destructive disease in southern coasts of the USA (Wilson and Gates 1993). For the first time blast disease in pearl millet was reported during 1952 at Government Research Farm, Kanpur, Uttar Pradesh, by Mehta et al. (1953). In India, blast on pearl millet has emerged as a serious threat (Lukose et al. 2007; Anonymous 2009), which becomes more destructive during humid weather conditions, especially when more plants per unit area are established. Earlier it was considered as a minor disease in India; presently pearl millet blast incidence has increased in several states of India and most predominantly on new commercial hybrids (Thakur et al. 2009). Blast disease has been occurring in all the states of major pearl millet-growing states of India since 1970 and its high incidence was observed recently in all the pearl millet-growing states like Gujarat, Madhya Pradesh, Uttar Pradesh, Delhi, Maharashtra, Rajasthan, and Karnataka (Nayaka et al. 2017).

Pyricularia setariae is the most destructive pathogen, which causes blast disease of foxtail millet. Under favorable conditions, the pathogen has the potential to cause yield losses up to 40% (Nagaraja 2007). It was first officially identified in Japan by Nishikado during 1917. In India it was reported from Tamil Nadu in 1919 (Mc Rae 1920). During 2016 blast disease outbreak on foxtail millet was observed for the first time in Mazandaran province of Iran (Pordel et al. 2018).

1.7 Disease Emergence and Spread of Wheat Blast

Recently, wheat blast disease is the most destructive disease caused by *M. oryzae Triticum* pathotype (MoT) and has the ability to cause complete crop failure under favorable weather conditions. The disease was first noticed in 1985 from Brazilian state of Parana (Igarashi et al. 1986). Later, it was reported from Santa Cruz Department of Bolivia in 1996 (Barea and Toledo 1996). The pathotype is distinct from the pathotypes infecting rice (the *Oryza* pathotype, MoO); finger millet (the *Eleusine* pathotype); Italian or foxtail millet (the *Setaria* pathotype); and turf grasses (the *Lolium* pathotype, MoL). During 2016, the disease has spread to Bangladesh which impacted around 15% of total wheat-growing area with an average yield loss of 51% in affected fields (Cruz and Valent 2017). This large-scale incidence outside South America has been a concern for the potential spread to other wheat-producing areas in Bangladesh, South Asia, and beyond. Alarmed with this the Indian Government has declared wheat holiday in the neighboring districts of West Bengal.

1.8 Distribution Pattern of Blast Disease

The severity and extent of damage caused by blast disease vary every year and from place to place depending on the weather conditions and early inception of disease. However, the detailed information collected on the blast endemic districts/rice-growing areas of the country is given in Table 1.1. The finger millet blast-occurring

Table 1.1 Occurrence of blast disease during rice cropping season in India

State	Rice ecology	Endemic districts/rice-growing area	Conducive period
Andhra Pradesh	Irrigated	Srikakulam, Vishakhapatnam, Guntur, Nellore, Chittoor, East and West Godavari	September–February
Assam	Irrigated	Karimganj, Tinsukia, Nowgong, Kamrup, Goalpara, North Lakhimpur	August–October
Bihar	Irrigated	Rohtas, Samastipur, Madhubani	August–October
Chhattisgarh	Irrigated	Jagdalpur, Northern hill regions	September–October
Gujarat	Irrigated	Kheda, Nawagam, Navasari	September–October
Haryana	Irrigated	Hissar, Karnal, Panipat, Sonipat, Kaithal	August–October
Himachal Pradesh	Irrigated/ Upland	Kangra Valley (Malan, Palampur), Kulu, Mandi	August–October
Jammu and Kashmir	Irrigated/ Upland	Hill zones of Anantnag, Rajouri, Jammu, Udhampur, Larnoo	July–September
Jharkhand	Irrigated	Ranchi, Hazaribagh	August–October
Karnataka	Irrigated	Mandya, Kodagu, Shimoga, Dharwad, Gangavathi	September–October
Kerala	Irrigated	Palghat, Kuttanad, Pattambi	September–February
Madhya Pradesh	Irrigated	Bastar region, Rewa, Bilaspur	September–October
Maharashtra	Irrigated	Pune, Ratnagiri, Kolaba, Parbhani, Kolhapur, Karjat	September–October
Manipur	Irrigated/ Upland	Wangbal, Manipur Central valley	July–October
Meghalaya	Irrigated/ Upland	West Khasi hills, Umiam, Upper Shillong	July–October
Mizoram	Irrigated/ Upland	Mizoram	August–October
Odisha	Irrigated	Cuttack, Ganjam, Koraput	July–August
Punjab	Irrigated	Amritsar, Bhatinda, Patiala, Ferozpur, Ropar, Hoshiarpur	August–October
Sikkim	Irrigated/ Upland	Gangtok	July–September
Tamil Nadu	Irrigated	Thanjavur, Coimbatore, Chengalpattu, South and North Arcot, Periyar, Madurai, Pudukkottai, Tirunelveli	October–February
Telangana	Irrigated	Rajendra Nagar, Jagtial, Ranga Reddy, Nizamabad, Medak, Mahbubnagar	September–February
Tripura	Irrigated/ Upland	West and South Tripura	July–October
Uttarakhand	Irrigated/ Upland	Almora, Bhageshwar, Dehradun, Nainital and other hill areas	August–October
Uttar Pradesh	Irrigated	Faizabad, Balia, Mathura, Meerut	August–October
West Bengal	Irrigated/ Upland	Darjeeling, Cooch Behar, Bankura	August–September

Table modified from source: Forewarning Rice Blast in India, Technical Bulletin No. 9, 2004–2005

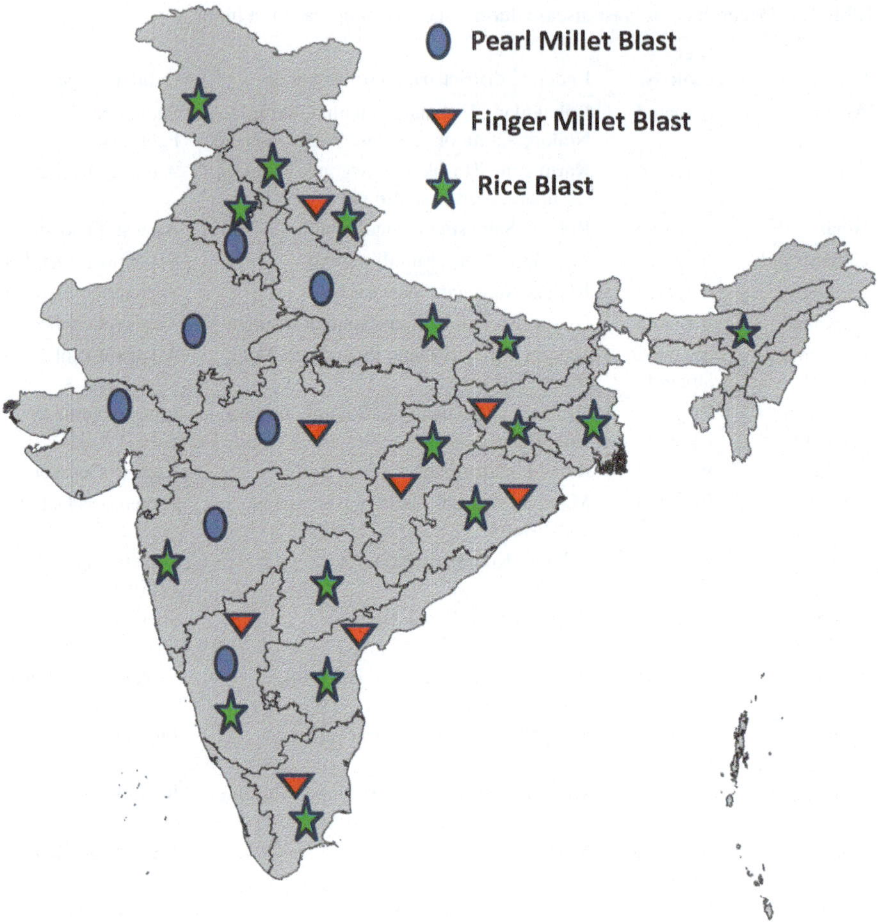

Fig. 1.4 Major blast hot spot locations of rice, pearl millet, and finger millet in India

states of India are Andhra Pradesh, Maharashtra, Karnataka, Tamil Nadu, Odisha, Uttarakhand, Madhya Pradesh, Bihar, and Chhattisgarh. The different pearl millet blast-occurring sates are also depicted (Fig. 1.4).

1.9 Establishment and Multiplication of Pathogen Isolates

Despite several decades of extensive studies on blast pathogen, the worldwide researchers always face difficulties in isolation of the pathogen from infected samples (leaf and neck) and its further establishment as monoconidial isolates (Jia 2009). Although several researchers proposed various isolation methods, the most commonly used was the slide moist chamber technique (Divya et al. 2013). In this method, the clear blast lesions are surface sterilized by washing with 0.1% mercuric chloride and with sterile double-distilled water three times. The sterilized sample is placed over a clean glass slide which is kept inside a sterile Petri dish padded with moist cotton. After incubation, a single spore from the sporulating lesions is identified using a stereomicroscope transferred aseptically to potato dextrose agar (PDA) plates. Besides PDA, studies have also reported the use of synthetic medium like Richards agar (Ramakrishnan 1948) and natural media like autoclaved leaves and grains of different hosts for multiplication of the pathogen. However, superior growth of the pathogen was obtained on rice leaf decoction (Nishikado 1927). Recent studies also reported that prune agar and oatmeal agar supported maximum mycelial growth and sporulation of the isolates both from rice and finger millet (Khadka et al. 2012). Similarly, stem bits of 20-day-old maize, rice, and *Panicum repens* are employed in mass multiplication of blast pathogen (Divya et al. 2013). Rajashekara et al. (2016) reported efficient methods for isolation and mass multiplication of blast isolates using spore drop technique.

1.10 Host Range and Cross-Infectivity Among Major Cereal Crops

The blast pathogen *M. grisea* (Cooke) Sacc. (Rossman et al. 1990) belongs to ascomycetes group of fungi; it is a heterothallic, filamentous fungus, infecting important cereal crops like rice, wheat, barley, and millets. In addition, the host range of the pathogen spans to almost 50 plant species belonging to 30 genera of Poaceae family economically (Ou 1985). Rice blast, caused by *Magnaporthe oryzae* (Ana. *Pyricularia oryzae*), is one of the most widespread and destructive diseases. The pathogen also infects wheat and other small grain crops (Valent and Chumley 1991; Tablot 2003). In India, rice blast pathogen has been reported to infect weed hosts like crab grass (*Digitaria sanguinalis*), *Eleusine coracana*, and *E. indica* (Singh 1997). Different host range studies of *Magnaporthe* sp. are determined by mass multiplication and artificial inoculation of the pathogen on different host plants under epiphytotic conditions. Several pathogenicity tests for establishing cross-infectivity between rice and crabgrass isolates are neither consistent nor comprehensive (Choi et al. 2013).

Infectivity of blast pathogen is chiefly restricted to its host species (Ramakrishnan 1948; Todman et al. 1994), although the cross infection between the plant species is established under artificial inoculations. In Uganda, blast isolates from weed species are able to establish on finger millet seedlings. This indicates that weeds/wild grasses act as "green bridges" for finger millet blast pathogen (Ekwamu 1988). This also suggests that different weed hosts growing on field bunds can serve as chief sources of primary inoculum, thereby initiating the disease development (Mackill and Bonman 1986). Hamer et al. (1989) and Valent et al. (1986) reported that the blast pathogen is strongly delimited by host range although it is able to infect a wide range of hosts. Inoculations of rice seedlings under artificial epiphytotic conditions with pathogen isolates of weeds resulted in successful (Mackill and Bonman 1986) and unsuccessful (Prabhu et al. 1992) cross-inoculations. Kulkarni and Govindu (1977) revealed interhost infection of blast pathogen between finger millet and foxtail, whereas both isolates did not infect rice seedlings. Viji et al. (2000) concluded that ten Indian isolates of blast pathogen from rice did not infect finger millet and vice versa, thereby confirming that the pathogen populations in India are distinct from each other. Similar findings are also reported by Kato et al. (1977) and Todman et al. (1994), whereas Kumar and Singh (1995) reported contradictory findings which may be due to the different environmental conditions prevailing during the time of experimentation and the status of nutrients in the soil (Asuyama 1965; Ou 1985). It is clear from different findings that the gene flow between rice and finger millet pathogen is majorly restricted to their respective host origin and they are considered as genetically distinct populations. The pathogenicity tests revealed that the isolates from infected weeds were pathogenic to finger millet and importantly some weed isolates are more aggressive than the finger millet isolates (Takan et al. 2004).

Pearl millet blast pathogen infects *Pennisetum glaucum*, *Pennisetum squamulatum*, *Pennisetum pedicellatum*, *Pennisetum macroforum* (Saikai et al. 1983), *Pennisetum purpureum* (Buckley and Allen 1951), and *Pennisetum ciliare* (Perrott and Chakraborty 1999). The pathogen also survives on other graminaceous hosts such as *Agrotis palustris*, *Brachiaria mutica*, *Cyperus rotundus*, *Eleusine indica*, *Eragrostis* sp., and *Panicum miliaceum* (Lanoiselet and Cother 2005). Isolates from *Panicum typhoides* did not infect rice, nor did *Pyricularia oryzae* (*Magnaporthe grisea*) attack bajra (Mehta et al. 1953).

Blast disease on wheat was first recorded in 1985 in the state of Parana, Brazil (Igarashi et al. 1986). Prabhu et al. (1992) showed that all the *P. grisea* isolates from rice, wheat, and grass weeds were pathogenic on wheat cultivars and barley in Brazil (Park et al. 2009). A special mechanism was reported to exist in these isolates so that the pathogen can infect Arabidopsis seedlings which are distinct from that of rice crop. Earlier, wheat blast outbreaks were limited and majorly reported in South America (Valent and Chumley 1991). Recently, wheat blast epidemics occurred in Bangladesh in the year 2016 (Cruz and Valent 2017).

1.11 Conclusions and Future Prospects

Blast disease of cereal crops is most widespread and destructive in nature and occurs virtually throughout the world. The pathogen is highly variable and it will vary from field to field and plant to plant during a particular season. The major findings of the studies showed that under artificial conditions mostly pathogen from one host could be able to infect another crop and *vice versa.* The continuous monitoring of pathogen shift from host to host is highly helpful for designing best management options for control of the disease.

References

Adipala E. Seed-borne fungi of finger millet. East Afr Agric Forest J. 1992;57:173–6.

Anonymous. Annual report, all India coordinated pearl millet improvement. Indian Council of Agricultural Research. Project Coordinating Unit, Agricultural Research Station, Mandore, Jodhpur. 2009.

Asuyama H. Morphology, taxonomy, host range and life cycle of *Pyricularia, oryzae*. In: Rice blast disease. Symposium Proceedings. Baltimore: The Johns Hopkins Press; 1965.

Babu TK, Thakur RP, Upadhyaya HD, Reddy PN, Sharma R, Girish AG, Sharma RK. Resistance to blast (*M. grisea*) in a mini-core collection of finger millet germplasm. Eur J Plant Pathol. 2013;135:299–311.

Barea G, Toledo J. Identificación y zonificación de piricularia o bruzone (*Pyricularia oryzae*) en el cultivo del trigo en el dpto. De Santa Cruz. CIAT. Informe Técnico. Proyecto de Investigación Trigo, Santa Cruz. 1996. p. 76–86.

Bisht IS. Blast tolerance and yield loss in finger millet. Indian J Agric Sci. 1987;57:954–5.

Buckley TA, Allen BF. Notes on current investigations. Malays Agric J. 1951;34:133–41.

Burnett F. Report on agriculture in Malaya for the year 1947. 1949. p. 87.

Cavara F. Contributzion ealla micologialombarda. Atti Instituto Botanico Universita Laboratorio Crittogamico Pavia Ser. 1892;2:207–92.

Choi J, Park SY, Kim BR, Roh JH, Oh IS. Comparative analysis of pathogenicity and phylogenetic relationship in *Magnaporthe grisea* species complex. PLoS One. 2013;8(2):e57196. https://doi.org/10.1371/journal.pone.0057196.

Cruz CD, Valent B. Wheat blast disease: danger on the move. Trop Plant Pathol. 2017;42:210–22.

Divya B, Robin S, Rabindran R, Manjunath H, Valarmathi P, Joel JH. Resistance reaction of genes introgressed lines against rice blast (*P. oryzae*) disease. Australas Plant Pathol. 2013;43:177–91.

Dunbar AR. The annual crop loss of Uganda. East African Literature Bureau, Dar es Salaam, Nairobi, Kampala. 1969.

Ekwamu A. Disease resistance of finger millet in Uganda. In: Proceedings of 6th EARSAM Regional workshop on sorghum and millet improvement in Eastern Africa, Mogadisha, Somalia. 1988. p. 20–7.

Fakrudin B, Shashidhar HE, Kulkarni RS, Hittalmani S. Genetic diversity assessment of finger millet, (*Eleusine coracana* (L) Gaertn.), germplasm through RAPD analysis. PGR Newsl. 2004;138:50–4.

Goto K. History of the blast disease and changes in methods of control. Agricultural Improvement Bureau, Ministry of Agriculture and Forestry, Japan. 1955;5:1–2.

Goulart ACP, Sousa PG, Urashima AS. Damages in wheat caused by infection of *Pyricularia grisea*. Summa Phytopathol. 2007;33:358–63.

Hamer JE, Farrall L, Orbach MJ, Valent B, Chumley FG. Host species specific conservation of family repeated DNA sequence in the genome of a fungal plant pathogen. Proc Natl Acad Sci. 1989;86:9981–5.

Herdt RW. Research priorities for rice biotechnology, in rice biotechnology. London: Alden Press Ltd; 1991. p. 35–7.

Igarashi S, Utiamada CM, Kasuma AH, Lopez RS. *Pyricularia* sp em trigo. 1. Ocurrencia de *Pyricularia* sp no Estado do Parana. Fitopatol Bras. 1986;11:351–2.

Jia Y. A user-friendly method to isolate and single spore the fungi *M. oryzae* and *M. grisea* obtained from diseased field samples. Online plant health progress. 2009. https://doi.org/10.1094/PHP-2009-1215-01-BR.

Kato H, Yamaguchi T, Nishihara N. Seed transmission pathogenicity and control of ragi blast fungus and susceptibility of ragi to *Pyricularia* spp. from grasses, cereals and mioga. Ann Phytopathol Soc Jpn. 1977;43:392–401.

Khadka RB, Shrestha SM, Manandhar HK, Gopal BKC. Study on differential response of *Pyricularia grisea* isolates from rice, finger millet and *Panicum* sp. with local and alien media, and their host range. Nepal J Sci Technol. 2012;13(2):7–14.

Kulkarni S, Govindu HC. Studies on the blast disease of ragi in Karnataka. IV. Host range of *Pyricularia setariae* Nishikado. Curr Res. 1977;6:30–1.

Kumar A, Singh RA. Differential response of *Pyricularia grisea* isolates from rice, finger millet and pearl millet to media, temperature, pH and light. Indian J Mycol Plant Pathol. 1995;25(3):238–42.

Kuwite CA, Shao FM. *Pyricularia* spp causing head blight of finger millet (*Eleusine coracana*) and other fungi associated with finger millet in Tanzania. In: de Millians WAJ, Frederiksen RA, Bengston GD, editors. Sorghum and millets diseases: a second world review. Patancheru: ICRISAT; 1992.

Lanoiselet V, Cother E. Plant Health Australia diagnostic protocol for *Magnaporthe grisea*. 2005. https://www.ippc.int/en/core-activities/capacitydevelopment/guides-and-training-materials/contributed-resource-detail/diagnostic-protocol-rice-blast-magnaporthe-oryzae/.

Lukose CM, Kadvani DL, Dangaria CJ. Efficacy of fungicides in controlling blast disease of pearl millet. Indian Phytopathol. 2007;60:68–71.

Mackill AO, Bonman JM. New hosts of *Pyricularia oryzae*. Plant Dis. 1986;70:125–7.

Manibhushanrao K. Rice blast disease. 1st ed. Delhi: Daya Publishing House; 1994. p. 1.

Mc Rae W. Detailed administration report of the government mycologist for the year 1919-20. 1920. Madras Agri. Dept.

Mehta PR, Singh B, Mathur SC. A new leaf spot disease of bajra (*Pennisetum typhoides* Staph and Hubbard) caused by a species of *Pyricularia*. Indian Phytopathol. 1953;5:140–3.

Mohamed MI. A disease of *Eleusine coracana* new to Somalia. Rev Agric Sutro Trop. 1980;74:73–9.

Muyanga S, Danial DL. Production and research review of small millets cereals in Zambia. In: Daniel DL, editor. Breeding for disease resistance with emphasizing durability. Wageningen: Wageningen Agril. Univ.; 1995. p. 60–4.

Nagaraja A. Compendium of small millets diseases. Project coordinator cell, all India coordinated small millets improvement project, UAS, GKVK Campus, Bengaluru, India. 2007. p. 80.

Nagaraja A, Nanja RYA, Anjaneya RB, Patro TSSK, Kumar B, Kumar J, Krishne GKT. Reaction of finger millet recombinant inbred lines (RILs) to blast. Crop Res (Hissar). 2010;39(1, 2 & 3):120–2.

Nayaka CS, Srivastava RK, Udayashankar AC, Lavanya SN, Prakash G, Bishnoi HR, Kadvani DL, Singh OV, Niranjana SR, Prakash HS, Satyavathi C. Magnaporthe blast of pearl millet in India-present status and future prospects. All India co-ordinated research project on pearl millet. Jodhpur: Mandor; 2017. p. 51.

Nishikado Y. Studies on rice blast disease. Jpn J Bota. 1927;3:239–44.

Ou SH. Blast in: rice diseases. 2nd ed. Kew: The Commonwealth Mycological Institute; 1985. p. 109–201.

Padmanabhan SY. Recent advances in the study of blast disease of rice. Madras Agric J. 1965;(Golden Jubilee Number):564–83.

Padmanabhan SY. Fungal diseases of rice in India. 1st ed. New Delhi: Indian Council of Agriculture Research; 1974. p. 15.

Park M. Report of the work done of the mycological division. Admn Rep Dir Agric Ceylon. 1932:103–11.

Park JY, Jin J, Lee Y, Kang S, Lee Y. Rice blast fungus (*M. oryzae*) infects Arabidopsis via a mechanism distinct from that required for the infection of rice. Plant Physiol. 2009;149:474–86.

Perrott RF, Chakraborty. *Pyricularia grisea* causes blight of buffel grass (*Cenchrus ciliaris*) in Queensland, Australia. Trop Grasslands. 1999;33:201–6.

Piotti E, Rigano MM, Rodino D, Rodolfi M, Castiglione S, Picco AM, Sala F. Genetic structure of *P. grisea* (Cooke) Sacc. isolates from Italian paddy fields. J Phytopathol. 2005;153:80–6.

Pordel A, Tharreau D, Cros-Arteil S, Shams E, Moumeni A, Mirzadi GA, Javan NM. *Pyricularia oryzae* causing blast on foxtail millet in Iran. Plant Dis. 2018;102(9):1853.

Prabhu AS, Filippi MC, Castro N. Pathogenic variation among isolates of *Pyricularia grisea* infecting rice, wheat, and grasses in Brazil. Trop Pest Manag. 1992;38:367–71.

Puri S, Kumar J. Characterization of leaf, neck and finger blast pathogen populations from *Eleusine coracana* and elucidating their interrelationships in *Eleusine-Magnaporthe* pathosystem. Indian Phytopath. 2012;65(2):133–41.

Rajashekara H, Prakash G, Pandian RTP, Sarkel S, Dubey A, Sharma P, Chowdary V, Mishra D, Sharma TR, Singh UD. An efficient technique for isolation and mass multiplication of *M. oryzae* from blast infected samples. Indian Phytopathol. 2016;69(4s):68–71.

Ramakrishnan KV. Studies on morphology, physiology and parasitism of the genus *Pyricularia* in Madras. Proc Indian Acad Sci Sec B. 1948;27(6):174–93.

Rao ANS. Estimates of losses in finger millet (*Eleusine coracana*) due to blast disease (*Pyricularia grisea*). J Agric Sci. 1990;24:57–60.

Rios JA, Debona D, Duarte HSS, Rodrigues FA. Development and validation of a standard area diagram set to assess blast severity on wheat leaves. Eur J Plant Pathol. 2013;136:603–11.

Rossman AY, Howard RJ, Valent B. *Pyricularia grisea*, the correct name of the rice blast disease fungus. Mycologia. 1990;82:509–12.

Saikai UN, Puzari KC, Dutta PK. *Pennisetum pedicellatum*, a new host of *Pyricularia grisea*. J Mycol Plant Pathol. 1983;12:237.

Shanmugapackiam S, Raguchander T. Host range and cross infectivity of the Genus *M. grisea*. Int J Curr Microbiol App Sci. 2018;7(8):3950–5.

Sharma TR, Rai AK, Gupta SK, Vijayan J, Devanna B, Ray S. Rice blast management through host-plant resistance: retrospect and prospects. Agric Res. 2012;1:37–52.

Shirai M. Notes on plants collected in suruga, Totomi, yamato and kii. Bot Mag Tokyo. 1896;10:111–4.

Singh RA. Current status of rice blast in India and challenges ahead. Indian Phytopathol. 1997;50(2):186–91.

Skamnioti P, Gurr SJ. Against the grain: safeguarding rice from rice blast disease. Trends Biotechnol. 2009;27:141–50.

Sundararaman S. Administration report of the Government mycologist, Coimbatore, for 1926-27, department report, Madras Presidency for the official year 1926-27. 1927. p. 326–44.

Tablot NJ. On the trail of a cereal killer: exploring the biology of *Magnaporthe grisea*. Annu Rev Microbiol. 2003;57:177–202.

Takan JP, Akello B, Esele P, Manyasa EO, Obilana AB, Audi PO. Finger millet blast pathogen diversity and management in East Africa: a summary of project activities and outputs. Int Sorghum Millets Newsl. 2004;45:66–9.

Thakur RP, Sharma R, Rai KN, Gupta SK, Rao VP. Screening techniques and resistance sources for foliar blast in pearl millet. J SAT Agric Res. 2009;7:1–5.

Tharreau D, Fudal I, Andriantsimialona D, Santoso, Utami D, Fournier E, Lebrun MH, Notteghem JL. World population structure and migration of the rice blast fungus, *M. oryzae*. In: Wang GL, Valent B, editors. Advances in genetics, genomics and control of rice blast disease; 2009. p. 209–15.

Thomas KM. Some aspects of the control of blast disease of paddy. Madras Agric J. 1930;18:596–604.

Thompson A. Notes on plant diseases in 1940. Malay agric J. 1941;24:241–5.

Todman AK, Pawar DR, Joshi MS. Host reactions to finger millet (*Pyricularia grisea* Sacc). Mysore J Agric Sci. 1994;28:45–6.

Valent B, Chumley FG. Molecular genetic analysis of the rice blast fungus, *Magnaporthe grisea*. Annu Rev Phytopathol. 1991;29:443–67.

Valent B, Crawford MS, Weaver CG, Chumley FG. Genetic studies of fertility and pathogenicity in *Magnaporthe grisea*. Iowa State J Res. 1986;60:569–94.

Venkatarayan SV. Diseases of ragi. Mysore J Agric Sci. 1946;24:50–7.

Viji G, Gnanamanickam SS, Levy M. DNA polymorphisms of isolates of *Magnaporthe grisea* from India that are pathogenic to finger millet and rice. Mycol Res. 2000;104(2):161–7.

Vishwanath S, Sanne GS, Seetharam A, Shankare GBT. Reaction to blast disease of released and pre-released varieties of finger millet from different states. Millet Newsl. 1986;5:31.

Wilson JP, Gates RN. Forage yield losses in hybrid pearl millet due to leaf blight caused primarily by *Pyricularia grisea*. Phytopathology. 1993;83:739–43.

[illegible]
[illegible]
[illegible]
[illegible]
[illegible] Appl [illegible] [illegible]
Singh [illegible]
Singh [illegible] and challenges [illegible]
[illegible]
[illegible]
[illegible]
[illegible] Nat [illegible]
[illegible]
[illegible] [illegible] Plant [illegible] [illegible]
[illegible]
[illegible]
Vander[illegible] [illegible] Press [illegible] 1968 [illegible]
[illegible]
[illegible]

Chapter 2
Utilizing Host-Plant Resistance to Circumvent Blast Disease in Rice

A. Khanna, R. K. Ellur, S. Gopala Krishnan, K. K. Vinod, P. K. Bhowmick, M. Nagarajan, B. Haritha, and A. K. Singh

Contents

2.1 Introduction

Rice is the important staple food crop of the world (Seck et al. 2012) which suffices 76% of the calorific needs of South Asia (Fitzgerald et al. 2008). Globally, India stands at first position with 44 mha area under rice cultivation with a production of ~116 million tons of milled rice during 2017–2018 (www.agricoop.gov.in). Basmati is a specialty rice which is mainly grown in the Indo-Gangetic plains of India which include states of Punjab, Haryana, Delhi, Uttarakhand, Himachal Pradesh, Jammu and Kathua districts of Jammu and Kashmir, and parts of western Uttar Pradesh. Basmati rice is mainly grown in an area of 1.5 mha with a production of 5.16 mt during 2018–2019. The annual forex earning due to export of Basmati rice is Rs. 32,806 crores (www.apeda.gov.in). Basmati rice possesses unique quality traits

A. Khanna · R. K. Ellur (✉) · S. Gopala Krishnan · K. K. Vinod · P. K. Bhowmick
M. Nagarajan · B. Haritha · A. K. Singh
ICAR-Indian Agricultural Research Institute, New Delhi, India

S. C. Nayaka et al. (eds.), *Blast Disease of Cereal Crops*, Fungal Biology,
https://doi.org/10.1007/978-3-030-60585-8_2

such as extra-long slender grain, exceptionally long-cooked grain, and soft and fluffy texture of cooked rice with strong and appealing aroma (Singh 2000). Major Basmati rice varieties under cultivation are Pusa Basmati 1, Pusa Basmati 1121, Pusa Basmati 6, Pusa Basmati 1509, etc. However, most of these popular rice varieties are highly susceptible to various biotic stresses, of which rice blast is one of the most devastating diseases of rice which cause significant yield losses.

2.2 Pathogen

Rice blast caused by a fungi *Magnaporthe oryzae* (Hebert) (Couch and Kohn 2002) [anamorph *Pyricularia grisea* (Cooke) Saccardo] is considered one of the most devastating diseases of rice causing significant yield losses. Annually, 10–30% of yield losses occur due to rice blast (Talbot 2003). *M. oryzae* is a hemi-biotroph which infects and grows in living plant cells but kills the infected cells once spread onto the adjacent cells, although it grows biotrophically in rice roots (Wilson and Talbot 2009). The infection begins when a three-celled conidium lands on a host leaf and anchors to the leaf cuticle. *M. oryzae* causes the disease in a cyclic developmental process. It initially attaches to the host surface through its conidia after which the spore germination takes place which extends into a germ tube. The germ tube undergoes hooking and swelling at its tip which further differentiates into a unicellular, dome-shaped infection structure called the appressorium. In 6–10 h, the appressoria differentiates, making its cell wall melanized to restrict permeability allowing only water molecules to pass through (Howard et al. 1991a). The cytoplasm then flows from the conidia into the appressorium and as the disease progresses, the latter matures and the former gets empty and collapses (Braun and Howard 1994). Glycerol accumulation engenders cellular turgor as high as 8 MPa which breaches the plant cell (Howard et al. 1991b). Once inside the cell, *M. oryzae* forms bulbous invasive hypha which develops a specialized structure called biotrophic interfacial complex responsible for secreting virulence factors. The disease spreads through the plasmodesmata to the neighboring cells. At this stage, the small oval chlorotic lesions begin to appear which further turn necrotic and then coalesce.

2.3 Host Range

M. oryzae is known to infect a wide variety of monocotyledonous plants such as rice (*Oryza sativa*), pearl millet (*Pennisetum glaucum*), foxtail millet (*Setaria italica*), wheat (*Triticum aestivum*), and barley (*Hordeum vulgare* L) but not the dicotyledonous plants (Schulze-Lefert and Panstruga 2011). The fungus can attack several other species of grasses including annual, perennial ryegrass and turfgrasses. It is also known to infect Arabidopsis via a mechanism distinct from that in rice (Park et al. 2009).

2.4 Symptoms and Damage

The blast pathogen attacks all aerial parts including leaf blades, leaf sheaths, rachis, nodes, panicles, and glumes. Blast occurs mainly during the vegetative stage of the crop cycle, although the major injury occurs during its reproductive stage called neck blast. The epidemics occur generally at maximum tillering followed by a sharp decline and a very low disease severity from flowering until the end of the crop cycle. Yield losses associated to neck blast are much higher than yield losses associated to leaf blast in the tropical rice lowlands of Asia (Ghatak et al. 2013).

2.5 Blast Prevention and Management

Rice blast hampers crop yield massively. Various cultural practices including broad-spectrum seed treatment, split application of nitrogen, and burning-contaminated straw and stubbles are employed to prevent the disease. Chemical measures are widely used by spraying prescribed doses of fungicides such as edifenphos, dithiocarbamate, benomyl, carbendazim, tricyclazole, and pyroquilon (Kumar et al. 2013). However, the cultural practices are not sufficient and usage of chemical pesticides is not an eco-friendly and bio-safe measure to counteract the pathogen. Therefore, developing genetic resistance is the most potent and durable approach to manage rice blast.

2.6 Genes Governing Blast Resistance

Rice blast patho-system follows the gene-for-gene model of host pathogen interaction (Flor 1956), according to which the key to disease resistance lies in the recognition of avirulence proteins by the plant resistance gene products (Silue et al. 1992). More than 100 blast resistance genes and more than 500 QTLs governing blast resistance have been identified (Ashkani et al. 2015). Many blast R-genes have been clustered on chromosomes 6 and 11 (Tanweer et al. 2015). Further, 36 blast resistance genes, viz., *Pib, Pit, Pita, Pita2, Pi1, Pi2, Pi5, Pi9, pi21, Pi25, Pi35, Pi36, Pi37, Pi50, Pi54, Pi56, Pi63, Pi64, Pizt, Pik, Pikm, Pikp, Pikh, Pike, Piks, Pigm, Pid2, Pid3, Pid3A4, Pish, Pb1, Pia, Pii, Pi-CO39, Pi54of,* and *Pi54rh*, have been characterized (Table 2.1; Wang et al. 2017; Meng et al. 2020). The first blast resistance gene to be cloned was *Pib* in Japan (Wang et al. 1999), while the first blast resistance gene to be cloned in India was *Pi54* (Sharma et al. 2005).

Correspondingly, >40 *Avr* genes have been identified using map-based cloning strategy (Feng et al. 2007); among these ten *Avr* genes, namely *AvrPita* (Orbach et al. 2000), *ACE1* (Böhnert et al. 2004), *AvrPia, AvrPii, AvrPik* (Yoshida et al. 2009), *AvrPiz-t* (Li et al. 2009), *Avr1-CO39* (Zheng et al. 2011), *AvrPib* (Zhang et al. 2015), *AvrPi9* (Wu et al. 2015), and *AvrPi54* (Ray et al. 2016), have been characterized with their corresponding *R*-genes.

Table 2.1 List of cloned blast resistance genes in rice

Sl. no.	Gene	Chr	Domain	Reference
1	*Pi37*	1	NBS-LRR	Lin et al. 2007
2	*Pit*	1	CC-NBS-LRR	Hayashi and Yoshida (2009)
3	*Pish*	1	CC-NBS-LRR	Takahashi et al. (2010)
4	*Pi35*	1	NBS-LRR	Fukuoka et al. (2014)
5	*Pi64*	1	CC-NBS-LRR	Ma et al. (2015)
6	*Pib*	2	NBS-LRR	Wang et al. (1999)
7	*pi21*	4	CC-NBS-LRR	Fukuoka et al. (2009)
8	*Pi63*	4	NBS-LRR	Xu et al. (2014)
9	*Pi9*	6	NBS-LRR	Qu et al. (2006)
10	*Pi2*	6	NBS-LRR	Zhou et al. (2006)
11	*Piz-t*	6	NBS-LRR	Zhou et al. (2006)
12	*Pi-d2*	6	Lectin receptor	Chen et al. (2006)
13	*Pi-d3*	6	CC-NBS-LRR	Shang et al. (2009)
14	*Pid3-A4*	6	NBS-LRR	Lü et al. (2013)
15	*Pi25*	6	CC-NBS-LRR	Chen et al. (2011)
16	*Pi50*	6	NBS-LRR	Zhu et al. (2012); Su et al. (2015)
17	*Pigm*	6	NBS-LRR	Deng et al. (2017)
18	*Pi36*	8	CC-NBS-LRR	Liu et al. (2007)
19	*Pi5*	9	CC-NBS-LRR	Lee et al. (2009)
20	*Pii*	9	NBS-LRR	Takagi et al. (2013)
21	*Pi56*	9	NBS-LRR	Liu et al. (2013)
22	*Pb1*	11	CC-NBS-LRR	Hayashi et al. (2010)
23	*Pia*	11	CC-NBS-LRR	Okuyama et al. (2011)
24	*Pik*	11	CC-NBS-LRR	Zhai et al. (2011)
25	*Pik-p*	11	CC-NBS-LRR	Yuan et al. (2011)
26	*Pike*	11	CC-NBS-LRR	Chen et al. (2015)
27	*Piks*	11	CC-NBS-LRR	GenBank: AET36547.1, AET36548.1
28	*Pikm*	11	NBS-LRR	Ashikawa et al. (2008)
29	*Pil*	11	NBS-LRR	Hua et al. (2012)
30	*Pi54*	11	NBS-LRR	Sharma et al. (2005, 2010)
31	*Pi54rh*	11	CC-NBS-LRR	Das et al. (2012)
32	*Pi54of*	11	CC-NBS-LRR	Devanna et al. (2014)
33	*PiK-h*	11	CC-NBS-LRR	Zhai et al. (2014)
34	*Pi-CO39*	11	NBS-LRR	Cesari et al. (2013)
35	*Pi-ta*	12	NBS-LRR	Bryan et al. (2000)
36	*Pi-ta2*	12	Armadillo repeat domain	Meng et al. (2020)

Table modified from Wang et al. (2017)

2.7 Development of Differential Varieties

A differential set consisting of 24 monogenic lines with resistance genes *Pia, Pib, Pik, Pikh, Pikm, Pikp, Piks, Pish, Pit, Pita, Pita2, Piz, Piz5, Pizt, Pi1, Pii, Pi3, Pi5 (t), Pi7 (t), Pi9, Pi11 (t), Pi12 (t), Pi19 (t),* and *Pi20 (t)* in the genetic background of a highly susceptible, japonica rice variety LTH was developed to monitor and predict the evolution of new forms of blast races (Tsunematsu et al. 2000; Fukuta et al. 2004). Additionally, another set of 27 monogenic NILs carrying various blast resistance genes in the genetic background of CO39 was developed (Telebanco-Yanoria et al. 2011).

2.8 Molecular Mechanism Underlying Blast Resistance via R-Genes

Molecular markers have become the most potent solution for overcoming blast. As the pathogen enters the host, specific avirulence protein (*Avr*) secreted by *M. oryzae* is recognized by the respective resistance (R) gene present in the resistant variety. The interaction leads to the activation of effector-triggered immunity (ETI) provoking hypersensitive reaction, thereby curbing the advancement of the pathogen cycle. The effector proteins secreted by the pathogen ignite the infection process to hijack the host machinery, which in turn are recognized by the R-gene to execute the defense mechanism pathway in the resistant varieties (Jain et al. 2017). Another defense mechanism involves the recognition of pathogen-associated molecular patterns (PAMP) by the plant pattern recognition receptors (PRR) activating pathogen-triggered immunity (PTI) (Dangl et al. 2013). However, the former mechanism imparts comparatively stronger and faster defense activation (Dong et al. 2013). The blast-resistant Pi genes encode the nucleotide-binding site-leucine-rich repeat (NBS-LRR) proteins and mostly impart ETI. The Pi genes are reported mostly to be constitutively expressed except *Pi5*-1 (Lee et al. 2009), *pi21* (Fukuoka et al. 2009), *Pi63* (Xu et al. 2014), and *Pb1* (Hayashi et al. 2010).

2.9 Marker-Assisted Backcross Breeding (MABB)

MABB is the promising approach to utilize molecular markers for incorporating target gene(s) into the elite crop variety without hampering its genetic background. A typical MABB comprises foreground, recombinant, and background selection. MABB requires the markers based on/linked to the target gene for foreground selection (Tanksley 1983), flanking markers for recombinant selection (Collard and Mackill 2008) and markers polymorphic between the recurrent and donor parents uniformly spanning across the genome for background selection (Hospital and Charcosset 1997).

With the availability of molecular markers linked to the blast resistance genes as well as genome-wide markers, MABB has been effectively utilized to improve rice varieties for blast resistance.

2.9.1 MABB for Developing Blast Resistance in India

The success of MABB has been depicted by various research programs utilizing the strategy with modified approaches. Distinct combinations of genes/QTLs have been introgressed into the background of popular cultivated varieties to overcome the virulent pathogenic spectrum of blast prevalent in a region.

Hittalmani et al. (2000) pyramided three blast resistance genes, i.e., *Pi2, Pi1,* and *Pita,* through MAS and concluded that *Pi2* exhibits higher level of resistance owing to the broader resistance spectrum of this gene. However, this study did not yield a commercial product. A combination of constitutively expressed gene *Pi2* from a donor parent C101A51 and a pathogen-induced gene *Pi54* from a donor parent Tetep was used to develop resistance in the genetic background of a PRR78, a restorer parent of a popular Basmati quality rice hybrid Pusa RH10, through marker-assisted simultaneous but stepwise method of backcross breeding (Singh et al. 2012a, b, 2013). The recurrent parent genome recovery in the developed NILs ranged from 78.33 to 89.01%. Further, the cooking quality and agronomic performance of these introgression lines were at par to the recurrent parent PRR78. Also, in another study the blast resistance genes *Piz5* and *Pi1* were incorporated into the genetic background of PRR78 by employing marker-assisted foreground selection using SSR markers AP5659-5 and RM5926, respectively. These lines were forwarded up to BC_3F_5 and the recovery of RPG ranged from 85.4 to 92.1% in the derived lines.

The first attempt to develop blast resistance in Basmati rice varieties was made by Khanna et al. (2015a, b). Pusa Basmati 1, a widely cultivated popular Basmati rice variety, was found highly susceptible to blast disease. Therefore, seven blast resistance genes, namely *Pi1, Pi54, Pita, Pi2, Pi9, Pi5,* and *Pib*, were incorporated into the genetic background of Pusa Basmati 1 through MABB using the gene-based/linked markers RM224, RM206, YL87/155, AP4007, AP5659-5, C1454, and RM208, respectively. A cafeteria of 36 NILs comprising 14 monogenic, 16 two-gene pyramids, and 6 three-gene pyramids was developed. Among the genes tested, *Pi9* was found to be most effective followed by *Pi2* and *Pita*. Therefore, one of the Pusa Basmati 1 NILs carrying blast resistance gene *Pi9* was released for commercial cultivation. Additionally, the blast resistance genes *Pi2* and *Pi54* were incorporated into the genetic background of most popular Basmati rice varieties Pusa Basmati 1121 and Pusa Basmati 6 (Ellur et al. 2016). In an effort to generate multiple biotic stress tolerance Singh et al. (2012a, b) carried out introgression of blast resistance gene *Pi54* along with a major QTL *qSBR11-1* for sheath blight using foreground and background selection in improved Pusa Basmati 1. The linkage drag was reduced up to 4.25 mb and 1.80 mb around *Pi54* and *qSBR11-1*, respectively, with RPG recovery of up to 89.50%.

Similarly, MABB was effectively used to incorporate blast resistance genes *Pi54*, *Pi1*, and *Pita* in the genetic background of a short-grain aromatic rice variety Mushk Budji. The gene-linked markers, namely Pi54-MAS, RM224, and YL87/155, were used in foreground selection to incorporate the genes *Pi54*, *Pi1*, and *Pita*, respectively; STS markers helped to reduce linkage drag around the genes *Pi54*, *Pi1*, and *Pita* to 2.74, 4.60, and 2.03 Mb, respectively; and 50K SNP chip analysis for background analysis revealed more than 92% genome similarity to recurrent parent (Khan et al. 2018).

2.9.2 *Blast-Resistant Varieties Developed Through MABB in India*

Pusa 1612 It is the near-isogenic line carrying blast resistance genes *Pi2* and *Pi54* from donor parent C101A51 and Tetep, respectively, in the genetic background of high-yielding aromatic rice variety Pusa Sugandh 5 through marker-assisted simultaneous but stepwise method of backcross breeding. Pusa 1612 was released for commercial cultivation for the Basmati-growing regions of the country (Table 2.2).

Pusa Basmati 1637 The blast resistance gene *Pi9* was transferred from the donor parent IRBL9-W into the genetic background of Pusa Basmati 1 through MABB. The gene-linked marker AP5659-5 was used in foreground selection of blast resistance gene *Pi9* and background selection using 104 genome-wide polymorphic SSR markers revealed the recurrent parent genome recovery of 96.6% with the linkage drag of <1.5 mb around the gene *Pi9*. This variety has been released for commercial cultivation in the Basmati-growing regions of the country (Table 2.2).

Pusa Basmati 1609 It is the blast-resistant Basmati rice variety possessing resistance genes *Pi2* and *Pi54*. This variety was developed by intercrossing two breeding lines Pusa 1602 and Pusa 1603 carrying blast resistance genes *Pi2* and *Pi54*, respectively, in the genetic background of PRR78. The markers AP5930 and RM206 linked to the genes *Pi2* and *Pi54* were used in foreground selection. This variety has been released for commercial cultivation in the Basmati-growing regions of the country (Table 2.2).

Table 2.2 List of rice varieties developed through MABB in India

Sl. no.	Recurrent parent	Donor parent	Blast resistance genes transferred	Improved variety released
1	Pusa Sugandh 5	C101A51 and Tetep	*Pi2* and *Pi54*	Pusa 1612
2	Pusa Basmati 1	IRBL9-W	*Pi9*	Pusa Basmati 1637
3	PRR78	Pusa 1602 and Pusa 1603	*Pi2 and Pi54*	Pusa Basmati 1609
4	Samba Mahsuri	DHMASQ164-2b	*Pi1, Pi54, Pita*	Pusa Samba 1850

Pusa Samba 1850 It is a MAS-derived near-isogenic line of the mega-rice variety "BPT 5204" (Samba Mahsuri) possessing three blast genes, namely *Pi54, Pi1,* and *Pita.* These genes have been transferred from a doubled haploid line, DHMASQ164-2b. In a MABB program, foreground selection was carried out using the gene-based/linked markers Pi54MAS (*Pi54*), YL87/155 (*Pita*), YL87/183 (*Pita*), and RM1233 (*Pi1*); stringent phenotypic selection was carried out for agro-morphological, grain, and cooking qualities; and background selection was carried out using 43 polymorphic markers with genome-wide coverage. The recurrent parent genome recovery in Pusa Samba 1850 was estimated to be 86.05%. This variety has been released for the states of Chhattisgarh and Odisha (Table 2.2).

2.10 Conclusion and Future Prospects

With the identification of several blast resistance genes, marker-assisted backcross breeding has become one of the most successful approaches to develop blast resistance in the popular rice varieties. Till date, four blast-resistant rice varieties have been developed through marker-assisted selection and released for commercial cultivation. Incorporating blast resistance genes should become a routine breeding program. Adoption of blast-resistant rice varieties would significantly reduce the use of fungicides and thereby produce consumer-safe produce. Blast disease caused by *M. oryzae* is considered as one of the notorious diseases with population dynamics varying from location to location. Therefore, effectiveness of various genes should be tested across locations before deploying the blast resistance genes. The deployment plan needs to be developed to overcome the blast disease as well as reduce the evolution and spread of new races.

References

Ashikawa I, Hayashi N, Yamane H, Kanamori H, Wu J, Matsumoto T, Ono K, Yano M. Two adjacent nucleotide-binding site-leucine-rich repeat class genes are required to confer *Pikm*-specific rice blast resistance. Genetics. 2008;180:2267–76.

Ashkani S, Yusop MR, Shabanimofrad M, Azadi A, Ghasemzadeh A, Azizi P, Latif MA. Allele mining strategies: principles and utilisation for blast resistance genes in rice (*Oryza sativa* L.). Curr Issues Mol Biol. 2015;17:57–74.

Böhnert HU, Fudal I, Dioh W, Tharreau D, Notteghem JL, Lebrun MH. A putative polyketide synthase/peptide synthetase from *Magnaporthe grisea* signals pathogen attack to resistant rice. Plant Cell. 2004;16:2499–513.

Braun EJ, Howard RJ. Adhesion of fungal spores and germlings to host plant surfaces. Protoplasma. 1994;181(1–4):202–12.

Bryan GT, Wu KS, Farrall L, Jia Y, Hershey HP, McAdams SA, Faulk KN, Donaldson GK, Tarchini R, Valent B. A single amino acid difference distinguishes resistant and susceptible alleles of the rice blast resistance gene *Pi-ta*. Plant Cell. 2000;12:2033–46.

Cesari S, Thilliez G, Ribot C, Chalvon V, Michel C, Jauneau A, Rivas S, Alaux L, Kanzaki H, Okuyama Y, Fournier EJ, Tharreau D, Terauchi R, Kroja T. The rice resistance protein pair RGA4/RGA5 recognizes the *Magnaporthe oryzae* effectors *AVR-Pia* and *AVR1-CO39* by direct binding. Plant Cell. 2013;25:1463–81.

Chen X, Shang J, Chen D, Lei C, Zou Y, Zhai W, Liu G, Xu J, Ling Z, Cao G, Ma B, Wang Y, Zhao X, Li S, Zhu L. A B-lectin receptor kinase gene conferring rice blast resistance. Plant J. 2006;46:794–804.

Chen J, Shi YF, Liu W, Chai R, Fu Y, Zhuang J, Wu J. A *Pid3* allele from rice cultivar Gumei2 confers resistance to *Magnaporthe oryzae*. J Genet Genomics. 2011;38:209–16.

Chen J, Peng P, Tian J, He Y, Zhang L, Liu Z, Yin D, Zhang Z. *Pike,* a rice blast resistance allele consisting of two adjacent NBS-LRR genes, was identified as a novel allele at the *Pik* locus. Mol Breed. 2015;35:117.

Collard BCY, Mackill DJ. Marker-assisted selection: an approach for precision plant breeding in the 21st century. Philos Trans R Soc B Rev. 2008;363:557–72.

Couch BC, Kohn LM. A multilocus gene genealogy concordant with host preference indicates segregation of a new species, Magnaporthe oryzae, from M. grisea. Mycologia. 2002;94:683–93.

Dangl JL, Horvath DM, Staskawicz BJ. Pivoting the plant immune system from dissection to deployment. Science. 2013;341:746–51.

Das A, Soubam D, Singh PK, Thakur S, Singh NK, Sharma TR. A novel blast resistance gene, *Pi54rh* cloned from wild species of rice, *Oryza rhizomatis* confers broad spectrum resistance to *Magnaporthe oryzae*. Funct Integr Genomics. 2012;12:215–28.

Deng Y, Zhai K, Xie Z, Yang D, Zhu X, Liu J, Wang X, Qin P, Yang Y, Zhang G, Li Q, Zhang J, Wu S, Milazzo J, Mao B, Wang E, Xie H, Tharreau D, He Z. Epigenetic regulation of antagonistic receptors confers rice blast resistance with yield balance. Science. 2017;355:962–5.

Devanna NB, Vijayan J, Sharma TR. The blast resistance gene *Pi54of* cloned from *Oryza officinalis* interacts with *Avr-Pi54* through its novel non-LRR domains. PLoS One. 2014;9:e104840.

Dong TG, Ho BT, Yoder-Himes DR, Mekalanos JJ. Identification of T6SS-dependent effector and immunity proteins by *Tn-seq* in *Vibrio cholera*. Proc Natl Acad Sci U S A. 2013;110(7):2623–8.

Ellur RK, Khanna A, Yadav A, Pathania S, Rajashekara H, Singh VK, Krishnan SG, Bhowmick PK, Nagarajan M, Vinod KK, Prakash G. Improvement of Basmati rice varieties for resistance to blast and bacterial blight diseases using marker assisted backcross breeding. Plant Sci. 2016;242:330–41.

Feng S, Wang L, Ma J, Lin F, Pan Q. Genetic and physical mapping of *AvrPi7*, a novel avirulence gene of *Magnaporthe oryzae* using physical position-ready markers. Chin Sci Bull. 2007;52:903–11.

Fitzgerald MA, Sackville Hamilton NR, Calingacion MN, Verhoeven HA, Butardo VM. Is there a second fragrance gene in rice? Plant Biotechnol J. 2008;6:416–23.

Flor HH. The complementary genetic systems in flax rust. Adv Genet. 1956;8(29):54.

Fukuoka S, Saka N, Koga H, Ono K, Shimizu T, Ebana K, Hayashi N, Takahashi A, Hirochika H, Okuno K. Loss of function of a proline-containing protein confers durable disease resistance in rice. Science. 2009;325:998–1001.

Fukuoka S, Yamamoto SI, Mizobuchi R, Yamanouchi U, Ono K, Kitazawa N, Yasuda N, Fujita Y, Nguyen TT, Koizumi S, Sugimoto K, Matsumoto T, Yano M. Multiple functional polymorphisms in a single disease resistance gene in rice enhance durable resistance to blast. Sci Rep. 2014;4:1–7.

Fukuta Y, Araki E, Telebanco-Yanoria MJ, Ebron L, Mercado-Escueta D, Takai T, Khush GS. Identification of blast resistance gene, *Pish* in rice (*Oryza sativa* L.). In: Proc. Plant and Animal Genome XII Conference, San Diego, CA, 10–14 January. 2004. p. 401.

Ghatak A, Willocquet L, Savary S, Kumar J. Variability in aggressiveness of rice blast (*Magnaporthe oryzae*) isolates originating from rice leaves and necks: a case of pathogen specialization. PLoS One. 2013;8(6):e66180.

Hayashi K, Yoshida H. Refunctionalization of the ancient rice blast disease resistance gene *Pit* by the recruitment of a retrotransposon as a promoter. Plant J. 2009;57:413–25.

Hayashi N, Inoue H, Kato T, Funao T, Shirota M, Shimizu T, Kanamori H, Yamane H, Hayano-Saito Y, Matsumoto T. Durable panicle blast-resistance gene Pb1 encodes an atypical CC-NBS-LRR protein and was generated by acquiring a promoter through local genome duplication. Plant J. 2010;64:498–510.

Hittalmani S, Parco A, Mew TV, Zeigler RS, Huang N. Fine mapping and DNA marker-assisted pyramiding of the three major genes for blast resistance in rice. Theor Appl Genet. 2000;100(7):1121–8.

Hospital F, Charcosset A. Marker assisted introgression of quantitative trait loci. Genetics. 1997;147:1469–85.

Howard RJ, Bourett TM, Ferrari MA. In: Mendgen K, Lesemann DE, editors. Infection by *Magnaporthe grisea*: an in vitro analysis. Berlin: Springer; 1991a. p. 251–64.

Howard RJ, Ferrari MA, Roach DH, Money NP. Penetration of hard substrates by a fungus employing enormous turgor pressures. Proc Natl Acad Sci. 1991b;88:11281–4.

Hua L, Wu J, Chen C, Wu W, He X, Lin F, Wang L, Ashikawa I, Matsumoto T, Wang L, Pan Q. The isolation of *Pi1,* an allele at the *Pik* locus which confers broad spectrum resistance to rice blast. Theor Appl Genet. 2012;125:1047–55.

Jain P, Singh PK, Kapoor R, Khanna A, Solanke AU, Krishnan SG, Singh AK, Sharma V, Sharma TR. Understanding host-pathogen interactions with expression profiling of NILs carrying rice-blast resistance *Pi9* gene. Front Plant Sci. 2017;8:9.

Khan GH, Shikari AB, Vaishnavi R, Najeeb S, Padder BA, Bhat ZA, Parray GA, Bhat MA, Kumar R, Singh NK. Marker-assisted introgression of three dominant blast resistance genes into an aromatic rice cultivar Mushk Budji. Sci Rep. 2018;8(1):1–3.

Khanna A, Sharma V, Ellur RK, Shikari AB, Gopala Krishnan S, Singh UD, Prakash G, Sharma TR, Rathour R, Variar M, Prashanthi SK, Nagarajan M, Vinod KK, Bhowmick PK, Singh NK, Prabhu KV, Singh BD, Singh AK. Development and evaluation of near-isogenic lines for major blast resistance gene(s) in Basmati rice. Theor Appl Genet. 2015a;128(7):1243–59.

Khanna A, Sharma V, Ellur RK, Shikari AB, Krishnan SG, Singh UD, Prakash G, Sharma TR, Rathour R, Variar M, Prashanthi SK. Marker assisted pyramiding of major blast resistance genes Pi9 and Pita in the genetic background of an elite Basmati rice variety, Pusa Basmati 1. Indian J Genet. 2015b;75(4):417–25.

Kumar MKP, Gowda DKS, Moudgal R, Kumar NK, Gowda KTP, Vishwanath K. Impact on fungicides on rice production in India. In: Nita M, editor. Fungicides-showcases of integrated plant disease management from around the world. InTech; 2013. ISBN: 978-953-51-1130-6. https://doi.org/10.5772/51009. http://www.intechopen.com/books/fungicides-showcases-of-integrated-plant-disease-managementfrom-around-the-world/impact-of-fungicides-on-rice-production-in-india. Accessed 18 Aug 2014.

Lee S, Song M, Seo Y, Kim S, Ko S, Cao P, Suh J, Yi G, Roh J, Lee S, An G, Hahn TR, Wang GL, Ronald P, Jeon JS. Rice Pi5-mediated resistance to *Magnaporthe oryzae* requires the presence of two coiled-coil-nucleotide-binding-leucine-rich repeat genes. Genetics. 2009;181:1627–38.

Li W, Wang B, Wu J, Lu G, Hu Y, Zhang X, Zhang Z, Zhao Q, Feng Q, Zhang H, Wang Z, Wang G, Han B, Wang Z, Zhou B. The *Magnaporthe oryzae* avirulence gene *AvrPiz-t* encodes a predicted secreted protein that triggers the immunity in rice mediated by the blast resistance gene *Piz-t*. Mol Plant Microbe Interact. 2009;22:411–20.

Lin F, Chen S, Que Z, Wang L, Liu X, Pan Q. The blast resistance gene *Pi37* encodes a nucleotide binding site leucine-rich repeat protein and is a member of a resistance gene cluster on rice chromosome 1. Genetics. 2007;177:1871–80.

Liu X, Lin F, Wang L, Pan Q. The in silico map-based cloning of *Pi36,* a rice coiled-coil-nucleotide-binding site leucine-rich repeat gene that confers race-specific resistance to the blast fungus. Genetics. 2007;176:2541–9.

Liu Y, Liu B, Zhu X, Yang J, Bordeos A, Wang G, Leach JE, Leung H. Fine-mapping and molecular marker development for *Pi56*(*t*), a NBS-LRR gene conferring broad-spectrum resistance to *Magnaporthe oryzae* in rice. Theor Appl Genet. 2013;126:985–98.

Lü Q, Xu X, Shang J, Jiang G, Pang Z, Zhou Z, Wang J, Liu Y, Li T, Li X, Xu J, Cheng Z, Zhao X, Li S, Zhu L. Functional analysis of *Pid3-A4,* an ortholog of rice blast resistance gene *Pid3* revealed by allele mining in common wild rice. Phytopathology. 2013;103:594–9.

Ma J, Lei C, Xu X, Hao K, Wang J, Cheng Z, Ma X, Ma J, Zhou K, Zhang X, Guo X, Wu F, Lin Q, Wang C, Zhai H, Wang H, Wan J. *Pi64,* encoding a novel CC-NBS-LRR protein, confers resistance to leaf and neck blast in rice. Mol Plant-Microbe Interact. 2015;28:558–68.

Meng X, Xiao G, Telebanco-Yanoria MJ, Siazon PM, Padilla J, Opulencia R, Bigirimana J, Habarugira G, Wu J, Li M, Wang B. The broad-spectrum rice blast resistance (R) gene Pita2 encodes a novel R protein unique from Pita. Rice. 2020;13(1):1–5.

Okuyama Y, Kanzaki H, Abe A, Yoshida K, Tamiru M, Saitoh H, Fujibe T, Matsumura H, Shenton M, Galam DC, Undan J, Ito A, Sone T, Terauchi R. A multifaceted genomics approach allows the isolation of the rice *Pia*-blast resistance gene consisting of two adjacent NBS-LRR protein genes. Plant J. 2011;66:467–79.

Orbach MJ, Farrall L, Sweigard JA, Chumley FG, Valent B. A telomeric avirulence gene determines efficacy for the rice blast resistance gene *Pi-ta*. Plant Cell. 2000;12:2019–32.

Park JY, Jin J, Lee YW, Kang S, Lee YH. Rice blast fungus (*Magnaporthe oryzae*) infects Arabidopsis via a mechanism distinct from that required for the infection of rice. Plant Physiol. 2009;149:474–86.

Qu SH, Liu GF, Zhou B, Bellizzi M, Zeng LR, Dai LY, Han B, Wang GL. The broad-spectrum blast resistance gene *Pi9* encodes a nucleotide-binding site-leucine-rich repeat protein and is a member of a multigene family in rice. Genetics. 2006;172:1901–14.

Ray S, Singh PK, Gupta DK, Mahato AK, Sarkar C, Rathour R, Singh NK, Sharma TR. Analysis of *Magnaporthe oryzae* genome reveals a fungal effector, which is able to induce resistance response in transgenic rice line containing resistance gene, *Pi54*. Front Plant Sci. 2016;7:1140.

Schulze-Lefert P, Panstruga R. A molecular evolutionary concept connecting nonhost resistance, pathogen host range, and pathogen speciation. Trends Plant Sci. 2011;16:117–25.

Seck PA, Diagne A, Mohanty S, Wopereis MCS. Crops that feed the world. Food Sec. 2012;4:7–24.

Shang J, Tao Y, Chen X, Zou Y, Lei C, Wang J, Li X, Zhao X, Zhang M, Lu Z, Xu J, Cheng Z, Wan J, Zhu L. Identification of a new rice blast resistance gene, *Pid3,* by genome wide comparison of paired nucleotide-binding site-leucine-rich repeat genes and their pseudogene alleles between the two sequenced rice genomes. Genetics. 2009;182:1303–11.

Sharma TR, Madhav MS, Singh BK, Shanker P, Jana TK, Dalal V, Pandit A, Singh A, Gaikwad K, Upreti HC, Singh NK. High-resolution mapping, cloning and molecular characterization of the *Pi-kh* gene of rice, which confers resistance to *Magnaporthe grisea*. Mol Gen Genomics. 2005;274:569–78.

Sharma TR, Rai AK, Gupta SK, Singh NK. Broad-spectrum blast resistance gene *Pi-kh* cloned from rice line Tetep designated as *Pi54*. J Plant Biochem Biotechnol. 2010;19:87–9.

Silue D, Notteghem JL, Tharreau D. Evidence of a gene-for-gene relationship in the Oryza sativa-*Magnaporthe oryzae* pathosystem. Phytopathology. 1992;82:577–80.

Singh VP. In: Singh RK, Singh US, Khush GS, editors. Aromatic rices. New Delhi: Oxford and India Book House; 2000. p. 135–53.

Singh VK, Singh A, Singh SP, Ellur RK, Choudhary V, Sarkel S, Singh D, Gopala SK, Nagarajan M, Vinod KK, Singh UD, Rathore R, Prashanthi SK, Agrawal PK, Bhatt JC, Mohapatra T, Prabhu KV, Singh AK. Incorporation of blast resistance into "PRR78", an elite Basmati rice restorer line through marker assisted backcross breeding. Field Crop Res. 2012a;128:8–16.

Singh A, Singh VK, Singh SP, Pandian RT, Ellur RK, Singh D, Bhowmick PK, Gopala Krishnan S, Nagarajan M, Vinod KK, Singh UD. Molecular breeding for the development of multiple disease resistance in Basmati rice. AoB Plants. 2012b;1:2012.

Singh VK, Singh A, Singh SP, Ellur RK, Singh D, Gopala Krishnan S, Bhowmick PK, Nagarajan M, Vinod KK, Singh UD, Mohapatra T, Prabhu KV, Singh AK. Marker-assisted simultaneous but stepwise backcross breeding for pyramiding blast resistance genes Pi2 and Pi54 into an elite Basmati rice restorer line PRR78. Plant Breed. 2013;132(5):486–95.

Su J, Wang W, Han J, Chen S, Wang C, Zeng L, Feng A, Yang J, Zhou B, Zhu X. Functional divergence of duplicated genes results in a novel blast resistance gene *Pi50* at the *Pi2/9* locus. Theor Appl Genet. 2015;128:2213–25.

Takagi H, Uemura A, Yaegashi H, Tamiru M, Abe A, Mitsuoka C, Utsushi H, Natsume S, Kanzaki H, Matsumura H, Saitoh H, Yoshida K, Cano LM, Kamoun S, Terauchi R. MutMap-Gap: whole-genome resequencing of mutant F2 progeny bulk combined with *de novo* assembly of gap regions identifies the rice blast resistance gene *Pii*. New Phytol. 2013;200:276–83.

Takahashi A, Hayashi N, Miyao A, Hirochika H. Unique features of the rice blast resistance *Pish* locus revealed by large scale retrotransposon-tagging. BMC Plant Biol. 2010;10:175.

Talbot NJ. On the trail of a cereal killer: exploring the biology of *Magnaporthe grisea*. Annu Rev Microbiol. 2003;57:177–202.

Tanksley SD. Molecular markers in plant breeding. Plant Mol Biol Rep. 1983;1:1–3.

Tanweer F, Rafii M, Sijam K, Rahim H, Ahmed F, Latif M. Current advance methods for the identification of blast resistance genes in rice. C R Biol. 2015;338:321–34.

Telebanco-Yanoria MJ, Koide Y, Fukuta Y, Imbe T, Tsunematsu H, Kato H, Ebron LA, Nguyet TMN, Kobayashi N. A set of near-isogenic lines of Indica-type rice variety CO39 as differential varieties for blast resistance. Mol Breed. 2011;27:357–73.

Tsunematsu H, Yanoria MJT, Ebron LA, Hayashi N, Ando I, Kato H, Imbe T, Khush GS. Development of monogenic lines of rice for blast resistance. Breed Sci. 2000;50:229–34.

Wang Z, Yano M, Yamanouchi U, Iwamoto M, Monna L, Hayasaka H, Katayose Y, Sasaki T. The *Pib* gene for rice blast resistance belongs to the nucleotide binding and leucine-rich repeat class of plant disease resistance genes. Plant J. 1999;19:55–64.

Wang B, Ebbole DJ, Wang Z. The arms race between *Magnaporthe oryzae* and rice: diversity and interaction of Avr and R genes. J Integr Agric. 2017;16(12):2746–60.

Wilson RA, Talbot NJ. Under pressure: investigating the biology of plant infection by *Magnaporthe oryzae*. Nat Rev Microbiol. 2009;7(3):185–95.

Wu J, Kou Y, Bao J, Li Y, Tang M, Zhu X, Ponaya A, Xiao G, Li J, Li C, Song MY, Cumagun CJR, Deng Q, Lu G, Jeon JS, Naqvi NI, Zhou B. Comparative genomics identifies the *Magnaporthe oryzae* avirulence effector *AvrPi9* that triggers *Pi9*-mediated blast resistance in rice. New Phytol. 2015;206:1463–75.

Xu X, Hayashi N, Wang C, Fukuoka S, Kawasaki S, Takatsuji H, Jiang C. Rice blast resistance gene *Pikahei-1(t)*, a member of a resistance gene cluster on chromosome 4, encodes a nucleotide-binding site and leucine-rich repeat protein. Mol Breed. 2014;34:691–700.

Yoshida K, Saitoh H, Fujisawa S, Kanzaki H, Matsumura H, Yoshida K, Tosa Y, Chuma I, Takano Y, Win J, Kamoun S, Terauchi R. Association genetics reveals three novel avirulence genes from the rice blast fungal pathogen *Magnaporthe oryzae*. Plant Cell. 2009;21:1573–91.

Yuan B, Zhai C, Wang WJ, Zeng XS, Xu XK, Hu HQ, Lin F, Wang L, Pan QH. The *Pik-p* resistance to *Magnaporthe oryzae* in rice is mediated by a pair of closely linked CC-NBS-LRR genes. Theor Appl Genet. 2011;122:1017–28.

Zhai C, Lin F, Dong ZQ, He XY, Yuan B, Zeng XS, Wang L, Pan QH. The isolation and characterization of *Pik*, a rice blast resistance gene which emerged after rice domestication. New Phytol. 2011;189:321–34.

Zhai C, Zhang Y, Yao N, Lin F, Liu Z, Dong Z, Wang L, Pan Q. Function and interaction of the coupled genes responsible for *Pik-h* encoded rice blast resistance. PLoS One. 2014;9:e98067.

Zhang S, Wang L, Wu W, He L, Yang X, Pan Q. Function and evolution of *Magnaporthe oryzae* avirulence gene *AvrPib* responding to the rice blast resistance gene *Pib*. Sci Rep. 2015;5:11642.

Zheng Y, Zheng W, Lin F, Zhang Y, Yi Y, Wang B, Lu G, Wang Z, Wu W. *AVR1-CO39* is a predominant locus governing the broad avirulence of *Magnaporthe oryzae* 2539 on cultivated rice (*Oryza sativa* L.). Mol Plant-Microbe Interact. 2011;24:13–7.

Zhou B, Qu SH, Liu GF, Dolan M, Sakai H, Lu GD, Bellizzi M, Wang GL. The eight amino-acid differences within three leucine-rich repeats between Pi2 and Piz-t resistance proteins determine the resistance specificity to *Magnaporthe grisea*. Mol Plant-Microbe Interact. 2006;19:1216–28.

Zhu X, Chen S, Yang J, Zhou S, Zeng L, Han J, Su J, Wang L, Pan Q. The identification of *Pi50(t)*, a new member of the rice blast resistance *Pi2/Pi9* multigene family. Theor Appl Genet. 2012;124:1295–304.

Chapter 3
Current Scenario and Integrated Approaches for Management of Finger Millet Blast (*Magnaporthe grisea*)

K. B. Palanna, Rajashekara Hosahatti, Gutha Venkata Ramesh, B. Malikarjuna, Boda Praveen, H. R. Raveendra, and T. S. S. K. Patro

Contents

K. B. Palanna (✉) · G. V. Ramesh · B. Malikarjuna · B. Praveen
Project Coordinating Unit, ICAR-AICRP on Small Millets, UAS, GKVK, Bengaluru, Karnataka, India

R. Hosahatti
ICAR-Vivekananda Parvatiya Krishi Anusandhan Sansthan, Almora, Uttarakhand, India

H. R. Raveendra
ICAR-AICRP on Small Millets, Zonal Agricultural Research Station, UAS (B), V.C. Farm, Mandya, Karnataka, India

T. S. S. K. Patro
Agricultural Research Station, ANGRAU, Gajularega, Vizianagaram, Andra Pradesh, India

S. C. Nayaka et al. (eds.), *Blast Disease of Cereal Crops*, Fungal Biology,
https://doi.org/10.1007/978-3-030-60585-8_3

3.1 Introduction

Finger millet [*Eleusine coracana* (L.) Gaertn.] originated in East Africa and is a nutrient-rich staple food for millions of people in the semiarid tropics of East Africa. It plays a crucial role in the subsistence farmer's economy and diets. It is world's fourth-ranked millet crop after sorghum, pearl millet, and foxtail millet. Finger millet generates sustainable income for millions of poor people in the semiarid regions of Eastern and Southern Africa, as well as South Asia. Finger millet is estimated to occupy 12 % of global millet area and accounts for cultivation in more than 25 countries in Africa and Asia. Uganda, India, Nepal, and China were stood as major producers. In the western Uganda and Ethiopian highlands, the crop was known to be domesticated around 5000 years BC and from there around 3000 BC, it reached to the west cost of India (Hilu et al. 1979). Finger millet is an essential food especially for the rural populations of Southern India and East and Central Africa. It can be grown under a wide range of climatic conditions ranging from plain areas to hilly regions of Himalayas but the crop performs better under well-drained, loamy soil. It is a highly productive crop that can survive well under various abrupt climatic conditions, and it can also grow as an organic crop. It has the capability to grow on low-fertile soils and is less dependent on the use of chemical fertilizers; hence, it is a boon for the huge arid and semiarid regions which can be grown by resource-poor farmers (Gull et al. 2014). Karnataka, Odisha, Maharashtra, Tamil Nadu, Andhra Pradesh, Uttarakhand, Uttar Pradesh, and Bihar are the major finger millet-growing states in India. It is cultivated at a wide range of altitudes in parts of Andhra Pradesh and Tamil Nadu to about 2400 m above sea level in hilly areas in northern India (Upadhyaya et al. 2007). Among the minor millet produce, finger millet constitutes about 81 % in India (Latha et al. 2005). About 60 % of finger millet is produced by the state of Karnataka which constitutes about 34 % of global production. Global warming has the major impact to cause a vicious cycle of disturbance in global ecosystems which have mostly negative effects on local natural fauna and flora. Impact of climate change has the potential to influence disease management aspects in various ways.

Finger millet is affected by various biotic and abiotic constraints. Among biotic constraints, blast disease caused by *Pyricularia grisea* Sacc. (teleomorph: *Magnaporthe grisea* (T. T. Hebert) M. E. Barr) is the prime catastrophic disease that causes substantial grain and forage yield losses. It is a serious problem in major finger millet-growing areas of Madhya Pradesh, Tamil Nadu. Karnataka, Kerala, and Maharashtra which causes heavy losses to the crop almost every year. It has also been recorded from other countries like Uganda, Tanganyika, and Malaya. The yield loss was reported by different workers from 50 to 100 % (McRae 1922; Venkatarayan 1946; Jegan et al. 2018). Rath and Mishra (1975) reported that neck infection causes great loss in grain number and grain weight and also spikelet sterility was increased significantly. Pall (1977) reported that neck infection resulted in considerable loss in panicle length, grain number, and grain weight. Management of blast disease is very challenging depending mainly on chemical fungicides like organophosphorus fungicides which are appreciatively effective (Kumar and Kumar 2011; Magar et al. 2015). However, excessive use of chemical fungicides has showed the development of pesticide-resistant fungal pathogens, with negative effects on the ecosystem like soil fertility and water quality and leading to dangerous health problems including birth defects (Hawkins et al. 2014; Hollomon 2016).

3.2 Distribution and Its Occurrence

Blast disease was spread to almost all the finger millet-growing regions of the world affecting different aerial parts of the plant starting from seedling till maturity. The disease is known to appear in India (Mc Rae 1920), Srilanka (Park 1932), Nepal (Thompson 1941), Malaya (Burnett 1949), Tanzania (Kuwite and Shao 1992), Somalia (Mohamed 1980), Zambia (Muyanga and Danial 1995), Ethiopia, Kenya, and Uganda (Dunbar 1969; Adipala 1992). In India the disease is prevalent in most of the finger millet-growing areas, viz. Karnataka, Tamil Nadu, Maharashtra, Andhra Pradesh, Orissa, Bihar, and Uttaranchal. The disease was first time reported in India from Tanjore delta of Tamil Nadu by Mc Rae (1920). It was subsequently reported from Karnataka (Venkatarayan 1937), Andhra Pradesh, Orissa, Bihar (Thirumalachar and Mishra 1953), Assam (Roy 1989), etc. Ramappa et al. (2002) registered up to 50 % neck blast and 70 % finger blast in Mandya and Mysore districts. The major blast hot-spot locations identified from India are shown in Table 3.1 (Anon 2018, 2019).

3.3 Pathogen Variability

The development of disease-resistant varieties/cultivars and knowledge on the pathogen population structure, such as the type of variants/haplotypes present in a location and the extent of variation, are very much essential for plant breeders to develop suitable resistant variety at a specific location. Therefore, specific delineation

Table 3.1 Major finger millet blast hot-spot locations of India

State	Hot-spot locations
Andhra Pradesh	Vizianagaram and Nandyal
Tamil Nadu	Athiyandal
Karnataka	Mandya and Bengaluru
Uttarakhand	Almora and Ranichauri
Chhattisgarh	Jagdalpur
Jharkhand	Ranchi
Odisha	Berhampur
Madhya Pradesh	Rewa

of pathogenic variability in the target production area is a primary factor for identifying finger millet genotypes with a stable resistance to the highly variable pathogen populations. From an ecological, epidemiological, and breeding perspective it is important to know how genetic diversity is maintained and how new, well-adapted highly virulent races evolve in the pathogen population. The frequent resistance breakdown mechanism of blast-resistant cultivars and studies on the extent of genetic diversity present in the population of *M. grisea* in a particular geographical region are effective (Levy et al. 1993). There is insufficient information available (Kumar et al. 2007) on the development of an uncertain set of differentials for assessing the racial differentiation for finger millet blast pathogen. Comprehensive work has been carried out with rice blast and detailed pathogenic variation has been described from single spores originating from single lesions and monoconidal subcultures (Ou and Ayad 1968; Ou et al. 1970).

Earlier evaluation of genetic diversity of *M. grisea* from various crops mostly relied on MGR-based restriction fragment length polymorphism (RFLP), which is a costly and time-consuming technique. The most commonly used DNA-based markers include randomly amplified polymorphic DNA (Williams et al. 1990; Welsh and McClelland 1990), amplified fragment length polymorphism (Vos et al. 1995), and sequence characterized amplified region (SCAR) markers (Soubabere et al. 2001). These markers are PCR based and any sequence information is not required; it is a speedy means to generate molecular markers but provides several genomic fragments with a marker in the single experiment (Varshney et al. 2007). However, these molecular markers are not locus specific, whereas RAPDs suffer with reproducibility. Microsatellites or SSR markers are random repeat DNA sequences present throughout the eukaryotic genome and on the other hand represent the locus-specific, highly polymorphic, multi-allelic, and codominant marker systems which have been proved to be the markers of choice in plant genetics and breeding applications (Gupta and Varshney 2000). Generation of SSR markers is a time-consuming, cumbersome, and expensive task. Several SSR (Brondani et al. 2000; Kim et al. 2000; Kaye et al. 2003; Suzuki et al. 2009) and minisatellite markers (Li et al. 2007) have already been developed for *M. grisea*. Dobinson et al. (1993) recognized a retro element in strains of *M. grisea* that infects finger millet and entitled it as grasshopper (*grh*). *M. grisea* isolates of rice and finger millet gathered from southern parts of India were characterized by MGR-DNA fingerprinting (Viji et al. 2000) and they also reported that the blast fungus did not cross-infect which was collected from these two hosts and also exhibited different fingerprint patterns. Takan et al. (2004) described that isolates causing leaf, neck, and panicle blast on finger millet compared by AFLP analysis were genetically similar indicating the same strains having the ability to cause different expressions of blast under suitable conditions. Degree of sexual compatibility that exists between rice and finger millet strains of *M. grisea* is high and there is great possibility of gene flow among these two host-limited populations of the pathogen (Rathour et al. 2004).

The population structure of rice, finger millet, jungle rice, goosegrass, and crabgrass infecting isolates of *M. grisea* from the north-western Himalayan region of India was analyzed using isozyme and native protein. Among different host-limited pathogen populations including those infecting rice and clustered in accordance with their host specificity, it showed a high level of genetic diversity. Subpopulations of the pathogen attacking rice and weeds were in the same field and were genetically distinct and there was no gene flow among rice and non-rice isolates of the pathogen (Rathour et al. 2006). Zheng et al. (2008), based on the released complete genome sequence data of *M. grisea*, developed polymorphic SSR markers (Dean et al. 2005) and constructed a genetic map consisting of 176 SSR markers. Sonah et al. (2009) noticed the high level of genetic variability through PCR-based RAPD analysis of *M. grisea* isolates from different non-rice and rice hosts and isolates from same location were grouped together irrespective of the crop from which infected samples were collected. Tanaka et al. (2009) examined the population structure of *Eleusine* isolates of *M. oryzae* by DNA fingerprinting with three repetitive elements, MGR586, MGR583, and *grasshopper*, which revealed that the isolates collected just after an outbreak of finger millet blast in Japan during 1970s had almost identical fingerprint profiles although they were collected in distant prefectures, and supported the idea that the outbreak was caused by seed transmission of a particular strain of *Eleusine* isolates. Takan et al. (2011) reported stable genetic variation pattern and lack of clonal lineages, with a broad range of haplotypes in 328 isolates of *M. grisea* from finger millet, rice, and *Dactylaria* spp. in East Africa.

3.4 Symptoms

The plant is susceptible at all growth stages starting from seedling till the time of grain formation. Based on the stage/plant parts affected the symptoms categorized three types, first symptom include leaf blast, as it is more severe in the tillering stage of the crop. The disease is diagnosed by spindle-shaped spots on the leaves with gray center surrounded by reddish brown margins as the disease progresses; there will be complete drying of foliage showing burnt appearance. The second symptom includes *neck blast, as it appears at the time of flowering* stage; the typical symptoms appear in the neck region just below the ear head which turns sooty black in color and usually breaks at the point of infection. In early neck infections, the entire ear head becomes chaffy and there is no grain formation. If grain setting occurs, they are shriveled and reduced in size. Neck infection results in great loss in grain number and grain weight and also spikelet sterility increases significantly (Rath and Mishra 1975). Third symptom represents *finger blast; it shows* infection on individual fingers during flowering stage, infected fingers turn black at the point of infection, the entire finger becomes chaffy, and there is no grain formation (Fig. 3.1). The finger-infected plants mature early compared to normal plants under field conditions.

Fig. 3.1 Blast symptoms and conidia of *M. grisea* (**a** leaf blast; **b** finger blast; **c** neck blast; **d**, **e** conidia)

3.5 Etiology

Blast is caused due to the Ascomycetes fungus *P. grisea* (Cooke.) Sacc. (formerly *P. oryzae* Cavara.) anamorph of *Magnaporthe grisea* (Hebert) Brar. It is a heterothallic, filamentous fungus pathogenic to almost 40 plant species belonging to 30 genera of Poaceae family (Ou 1980; Murakami et al. 2000; Inukai et al. 2006) including *Eleusine*. Initially, the opinion with regard to the nomenclature of the pathogen was different. The correct identity of the pathogen is still uncertain. Park (1932) reported *P. oryzae* on ragi in Uganda. In Malaya also *P. oryzae* has been recorded on this host. The isolate from ragi does not infect rice. Ramakrishnan (1948) recorded it to be a race of *P. oryzae*. Wallace and Wallace (1948) also reported *P. oryzae* from Tanganyika on ragi. Wallace and Wallace (1948) classified it to be as *P. setariae*. Morphologically it is very similar to *P. oryzae* (Ramakrishnan 1948). The perfect stage of *Pyricularia grisea* was earlier named as *Ceratosphaeria grisea* (Hebert 1971). Later Yaegashi and Nishihara (1976) suggested the genus *Magnaporthe*. Yaegashi and Udagawa (1978) finally gave *M. grisea* as the perfect stage of *P. grisea* (Cke.) Sacc. instead of *Ceratosphaeria grisea*. Chauhan and Varma (1981) reported *P. grisea* on *Eleusine indica* from Kanpur, India.

Young hyphae are hyaline and septate, and older hyphae are brown colored. Numerous conidiophores and conidia are formed in the middle portions of the lesions under humid conditions. The upper surface is darker than lower surface. The conidiophores emerge through epidermal cells or stomata. They are septate straight, subhyaline at the tip, and dark colored at the base. The conidia are hyaline, thin walled, and subpyriform. The size of the conidia varies, being 19–31μ × 10–15μ. Each pyriform spore is 2–3 celled, the middle cell being darker and broader than others (Fig. 3.1). They are formed acrogenously one after another, by the sympodial growth of the conidiophore. Conidiophores are simple and septate, with basal portion being comparatively darker. The pathogen produces fertile perithecia under laboratory conditions (Viji and Gnanamanikam 1998).

3.6 Mode of Spread and Survival of the Pathogen

The fungus enters the host tissue by piercing through the epidermal cells or through stomata. The incubation period varies from 4 to 6 days. Intensity of the disease depends on the weather conditions prevailing in a given season. The initial inoculum of pathogen comes from alternate hosts like weeds, collateral hosts, plant debris, and shriveled seeds. Kato and Nishihara (1977) also recorded the survival of fungus on seeds. The pathogen readily infects foxtail millet, bajra, ragi, wheat, barley, oats, maize, and crowfoot grass. Kato and Nishihara (1977) also reported that *Pyricularia* isolates from *E. coracana, E. indica, E. africana*, and *E. floccifolia* were pathogenic to ragi. Isolates from *E. coracana, E. indica, Setaria viridis* var. minor were pathogenic to *Lolium multiflorum, Festuca elatior* var. *arundinacea, Phalaris arundinacea, Anthoxanthum odoratum,* maize, barley and oats. One infected seed could be able to cause an epidemic (Pall 1988).

3.7 Yield Loses

Several workers have investigated on the adverse effect of blast disease on the yield of ragi; immediately after its first report in 1920, McRae (1922) recorded that the loss of grain may amount to over 50 % and subsequently Venkatarayan (1946) recorded 80–90 % loss in yield in erstwhile Mysore state (Venkatarayan 1946); elsewhere blast was observed to destroy more than 10 % ear heads (Anon 1959). Disease survey in Andhra Pradesh, Delhi, Haryana, Madhya Pradesh, Maharashtra, and Karnataka showed that blast causes a serious loss in finger millet crop (Sundaram et al. 1972). Neck infection causes significant loss in grain number and grain weight in different varieties accompanied by significant increase in spikelet sterility. Reduction in spikelet number is less consistent than other characteristics (Rath and Mishra 1975). Pall (1977) reported neck infection root to be the cause for considerable yield loss in panicle length, grain number, and grain weight. The mean yield loss is 46 % in infected ears compared to the healthy ones and there was a loss of 20.9 % in processing, resulting in an effective loss of 63.72 % (Rao 1981).

Rao and Hegde (1987) reported that when the average disease incidence was 7.69 % in the neck and 9.97 % in the finger, the loss due to blast was 29.51 %. In further studies, Rao (1990) observed that the loss in ragi ranged from 6.75 to 87.5 %. A rise of 1 % infection in the neck and finger resulted in a similar increase of 0.32 and 0.084 % in yield losses. In India, the average loss due to blast has been reported to be around 28–36 % (Vishwanath et al. 1986; Nagaraja et al. 2007), and in endemic areas, yield loss can be as high as 80–90 % (Vishwanath et al. 1986; Bisht 1987; Rao 1990). Ragi blast in Himalayan region appears at lower elevation and it was recorded at <1600 m and caused 25–40 % yield loss (Bisht et al. 1997). Cent percent yield reduction was recorded at Rampur, Nepal (Batsa and Tamang 1983). Grain yield losses associated to blast were predicted to be between 10 and 50 % in Kenya. In Uganda, blast incidence (13–50 %) and severity (24–68 %) varied significantly across main finger millet-cultivated areas in the North and East (Takan et al. 2004). Gupta (1997) reported a grain yield loss of 56 % due to blast. Vishwanath et al. (1997) reported the average annual loss due to blast in finger millet to be around 28 % with a range of 15–36 %. Ramappa et al. (2006) reported that different dates of sowing have showed different levels of yield loss. Dagnachew et al. (2014) reported 42 %; Prajapati et al. (2013) 35.78 %; Jegan et al. (2018) 50–100 %; Rao (1990) 6.75–87.5 %; and Kumar et al. (2005) up to 28 % colossal loss annually.

3.8 Disease Forecasting and Epidemiology

Forecasting of the disease occurrence is a very effective tool in the management of several economically important plant diseases. Knowledge on relationships between weather variables and blast disease could be used to strengthen techniques to screen for resistance and to design effective strategies for management of disease. For example, mist was used to provide high relative humidity and leaf wetness that are ideal for initial infection development which is already being used for screening pearl millet for blast resistance (Thakur et al. 2009). In general, favorable conditions for blast disease development were long periods of leaf wetness, high relative humidity, and temperature range of 17–28 °C. These factors are more appropriate with a polycyclic, airborne pathogen like *Pyricularia* spp.

No serious attempts have been made with regard to forecasting of blast of finger millet. Thomas (1940) observed high incidence of blast on crops sown in June, July, and August; less incidence was observed in May- and September-sown crops and negligible incidence in the crops sown in the remaining months. Fortnightly sowing of susceptible varieties, recording the onset of blast on each sowing, and relating it to weather variables are being done under All India Coordinated Research Project (AICRP) on small millets at Bengaluru center for the past several years. Blast incidence was high on crops sown in August and less severe on crops sown during second fortnight of June. The average minimum and maximum temperatures were 22 °C and 29 °C, respectively, with 85–99 % RH during the growth period (Patel and Tripathi 1998). Kumar et al. (2005) stated that increased neck and finger blast incidence was due to decreased temperature and increased relative humidity; opposite trends were recorded for low blast disease development. September and October months resulted in more than 50 % leaf blast severity in the nursery at 20 days after sowing. During these months, maximum rainfall, higher number of rainy days, high relative humidity, and low minimum temperature were recorded (Ramappa et al. 2006). Low temperature, high RH (>80 %), and high rainfall support blast development (Nagaraja et al. 2010).

The climatic conditions that were more favorable for blast disease development prevailed from first fortnight of July with an average minimum and maximum temperature of around 20 °C and 30 °C, respectively, and relative humidity of >80 % (Bisht et al. 1984). The temperature range of 18–24 °C was more congenial for the development of neck and finger blast in ragi than at other temperature ranges (Chaudhary and Vishwadhara 1988; Gowda and Gowda 1995; Kumar et al. 2005). Ramappa et al. (2006) reported that highest leaf blast severity over 50 % in finger millet was observed in nursery raised in October month probably due to availability of high inoculum pressure coincided with favorable weather conditions, i.e., more number of rainy days, high relative humidity, and low night temperature recorded during the month of October.

3.9 Mechanism of Resistance

The mechanism of resistance in plants towards blast disease is not clearly understood. Susceptibility to blast was found to be positively correlated with protein content and most of the high-yielding varieties were low in proteins (Dineshkumar et al. 1985). *Pyricularia* infection resulted in increase in protein content of seed and reduced starch and ash. β-Glucosidase activity was greater and glucose content lesser in disease portion of the neck than in healthy tissue (Pall 1992). There is a relationship between total phenols, total tannins, and level of susceptibility to blast fungus (Kumar and Singh 1995). In fact, the brown grain types are resistant to blast compared to white grain (Ravikumar and Seetharam 1993). The total protein and reducing sugar content was more in susceptible variety compared to that in resistant variety, and total phenol and tannin content was high in the resistant variety compared to that in susceptible variety (Somappa 1999). Byregowda et al. (1998) revealed that the resistance/susceptibility appears to be the result of multiple biochemical compounds present in plant and no single mechanism solely accounted for disease resistance. The resistant genotypes consistently had higher levels of phenols and tannins. Interestingly the attributes like phenol, tannin content, and grain yield showed high heritability and high genetic advance representing the role of additive genes (Byregowda et al. 1999a). The resistant genotypes were found to possess thicker leaves, significantly thicker upper epidermis, and a less pronounced reduction in thickness consequent to infection compared to susceptible varieties (Somappa 1999).

The resistance mechanism of leaf, neck, and finger blast has showed that less leaf area, narrow leaf angle, less number of stomata, short plant with better conversion efficiency of photosynthates from source to sink (harvest index), thick epidermis and cuticle on the leaf and neck, fewer chlorenchymatous strands, higher total phenols, and low quantities of total and reducing sugars contributed towards blast resistance in finger millet (Jain and Yadava 2004). Blast disease recorded low heritability and moderate genetic advances indicating the role of nonadditive gene effect (Byregowda et al. 1999b). Neck blast and finger blast were positively correlated with glume cover, seed protein content, and peduncle length and negatively correlated with seed calcium content, days to flowering, and yield, and there was no relationship between grain color and blast resistance (Nagaraja et al. 2010).

3.10 Integrated Disease Management

Ever since the first report on the distribution and existence of blast disease by Mc Rae in 1922, various workers are making concerted efforts from time to time to manage the disease. Many strategies were adopted like resistant genotypes coupled with good agronomic traits. Likewise, there are several highly effective fungicides and biocontrol agents were also identified for management of disease.

3.10.1 Cultural Methods

The incidence of blast is much high in direct-seeded ragi than in the transplanted crop. This may be due to thick plant population in direct seedling which alters the microclimate which is favorable to multiplication and rapid spread of the pathogen (Mishra et al. 1985). Application of increased levels of potassium has decreased blast severity while nitrogen application enhances the incidence of blast severity. Calcium silicate seems superior over sodium silicate in reducing neck blast and finger blast on the test genotypes (Krishnappa et al. 2013). Change in sowing time was useful in avoiding/escaping the incidence of neck and finger blast infection and in turn getting higher grain yields (Nagaraja et al. 2007). Seed treatment gave good control of leaf blast regardless of the cultivar used (Madhukeshwara et al. 2005).

3.10.2 Host Plant Resistance (HPR)

Exploiting host resistance to control disease is not only economical but also a practical necessity in a low-value crop like finger millet where there is a limitation for any additional cash inputs such as fungicides. Development of resistance varieties is the best means of combating the disease, which is predominantly grown by resource-poor and marginal farmers. The success of such programs depends on the identification of durable resistant sources and its subsequent utilization in breeding. The search still continues for sources of high levels of host-plant resistance (HPR).

However, large-scale evaluation of germplasm collections against various biotic or abiotic stresses is resource consuming and time consuming. Blast-resistant varieties identified and released for the different finger millet-growing areas of India are tabulated in Table 3.2 (www.aicrpsm.res.in).

3.10.3 Biological Control

Biological control is an alternative to synthetic chemical pesticides and has several benefits to human beings and ecosystem; they can ensure the protection of plants against biotic and abiotic stresses, ensure production of good-quality grains, improve soil fertility, and assure sustainable and safe environment. The demand for development and application of indigenous bioinoculant products has increased among researchers because of their role in plant growth promotion and crop protection in sustainable farming systems and also for their economic value (Schreiter et al. 2014; Santhanam et al. 2015; Sekar et al. 2016; Cai et al. 2017). However, the performance of bioinoculants, viz. *Pseudomonas, Trichoderma,* and *Bacillus*, in the field highly depends on their survival ability to express key traits in the soil without adversely affecting the native soil microbes (Gupta et al. 2015; Thomas and Sekhar 2016; Sharma et al. 2017).

There has been concerted effort to test various biocontrol agents and compounds in the management of blast of finger millet. *Gliocladium virens* and *Trichoderma viride* were tested as seed dressers to see their efficacy on blast. Both reduced the leaf blast significantly and were on par with standard seed dressing fungicide, carbendazim (Somappa 1999). *Pseudomonas* sp. (strain MSSRFD41) showed a 22.35 mm zone of inhibition against *P. grisea*; produced antifungal siderophores, metabolites, IAA, and hydrolytic enzymes; and solubilized phosphate. Environmental SEM analysis indicated the potential of MSSRFD41 to inhibit the growth of *P. grisea* by affecting cellular functions, which caused distortion in fungal hyphae. Bio-primed finger millet seeds showed significantly higher levels of germination and seedling vigor index and enhanced shoot and root length compared to check seeds. Cross streaking and RAPD analysis showed that MSSRFD41 is companionable with different sets of rhizobacteria and lived in the rhizosphere. In addition, PLFA analysis showed no significant variation in microbial biomass between the treated and control rhizosphere samples. Field trials showed that MSSRFD41 treatment significantly decreased blast infestation and improved plant growth compared to other treatments. A liquid-formulated MSSRFD41 product maintained shelf life at an average of 10^8 CFU mL^{-1} over 150 days of storage at 25 °C. Overall, results from this study revealed that *Pseudomonas* sp. MSSRFD41, an indigenous rhizobacterial strain, is an effective, alternative, and sustainable resource for the control of *P. grisea* infestation and growth promotion of finger millet (Jegan et al. 2018). Several workers have reported pseudomonas as bioinoculants with the potential to control phytopathogens and promote the growth of crops cultivated under varied agroclimatic conditions (Yin et al. 2013; Selvaraj et al. 2014; Wang et al. 2015; Jegan et al. 2018).

Table 3.2 Blast-resistant varieties identified and released for the different finger millet-growing areas of India (Source adopted from www.aicrpsm.res.in)

S no.	Variety	Pedigree	Developed institute	Year of release	Area of adaption
1	Saptagiri (PR 2614)	MR 1 × Kalyani	APAU, Perumallapalle	1995	Tamil Nadu, Maharashtra, Odisha, Andhra Pradesh
2	A 404	Selection from germplasm	BAU, Ranchi	1993	Bihar
3	Gautami (PR 1158-9)	PR 202 × U22	MRS, Vizianagaram	1993	Andhra Pradesh
4	Suraj (VR 520)	Pureline selection from VM 2507/19	ANGRAU, Vizianagaram (AP)	1994	All over India
5	KM 65	Selection from exotic germplasm	CSAUA&T, Kanpur	1994	Uttar Pradesh
6	BM 2	Pureline selection	BAU, Ranchi	1995	Bihar
7	GPU 28	Indaf 5 × (Indaf 9 × IE 1012)	PC Unit, Bangalore	1996	Karnataka
8	PR 230 (Maruthi)	Pureline selection	ANGRAU, Paleru	1998	Andhra Pradesh (Telangana region)
9	BM 9-1	Mutant from Budha Mandia	OUAT, Berhampur	1999	Karnataka, Andhra Pradesh, Odisha, Madhya Pradesh, Maharashtra
10	GPU 26	(I-5 × I-9) × IE 1012	PC Unit, UAS, Bengaluru	2000	Karnataka
11	GPU 26	(I-5 × I-9) × IE 1012	PC Unit, UAS, Bengaluru	2000	Karnataka
12	GPU 45	GPU 26 × L 5	PC Unit, UAS, Bengaluru	2001	Gujarat, Jharkhand, Karnataka, Madhya Pradesh, Maharashtra
13	Chilika (OEB 10)	GE 68 × GE 156	OUAT, Bhubaneshwar, Odisha	2001	Odisha, Madhya Pradesh, Gujarat, Andhra Pradesh, Tamil Nadu
14	VL 315	SDFM 69 × VL 231	VPKAS, Almora	2004	Uttaranchal
15	GPU 48	GPU 26 × L 5	PC Unit, UAS, Bengaluru	2005	Karnataka
16	GPU 48	GPU 26 × L 5	PC Unit, Bengaluru	2005	Karnataka
17	PRM 1	Selection from Ekeshwar of Pauri Garhwal Region	Hill Campus, GBPUA and T, Ranichauri	2006	Hills of Uttarakhand

(continued)

Table 3.2 (continued)

S no.	Variety	Pedigree	Developed institute	Year of release	Area of adaption
18	Bharathi (VR 762)	Pureline selection from VMEC 134	ANGRAU, Vizianagaram	2006	Andhra Pradesh
19	Srichaitanya (VR 847)	GPU 26 × L 5	ANGRAU, Vizianagaram	2009	Andhra Pradesh
20	KMR 301	MR 1 × GE 1409	VC Farm, Mandya, UAS, Bengaluru	2009	Southern dry zone of Karnataka
21	KOPN 235	Selection from local germplasm	MPKVV, Rahuri	2011	Sub-mountain and ghat zone of Maharashtra
22	OEB 526	SDFM 30 × PE 244	OAUT, Bhubaneswar	2011	Odisha, Bihar, Chhattisgarh, Karnataka, Tamil Nadu
23	OEB 532	GPU-26 × L-5	OAUT, Bhubaneswar	2012	Odisha, Bihar, Chhattisgarh, Karnataka, Tamil Nadu
24	PPR 2700 (Vakula)	KM 55 × U22/B	ARS, Perumallapalle, AP	2012	Andhra Pradesh
25	VL 352	VR 708 × VL-149	ICAR-VPKAS, Almora	2012	All ragi-growing areas of country
26	Chhattisgarh Ragi-2	PR 202 × GE 669	Jagdalpur, IGKVV	2012	Chhattisgarh
27	VL 376	GE 4172 × VL Ragi 149	ICAR-VPKAS, Almora	2016	All ragi-growing areas of country
28	GNN-6	Selection from local germplasm WN-259	Waghai, Navsari Agricultural University	2016	Gujarat
29	GN-5	Selection from local germplasm WWN-20	Waghai, Navsari Agricultural University	2016	Gujarat
30	VL Mandua-348	VL Ragi 146 × VL Ragi 149	ICAR-VPKAS, Almora	2016	Uttarakhand
31	KMR 340	OUAT-2 × WRT-4	VC Farm, Mandya, UAS, Bengaluru	2016	Karnataka
32	Dapoli-2 (SCN-6)	Somaclone of Dapoli-1	Dr. BSKKV, Dapoli	2017	Konkan region of Maharashtra
33	CO 15	CO 11 × PR 202	Centre on Excellence of Millets, TNAU, Athiyandal, Tamil Nadu	2017	Tamil Nadu

The influence of bioinoculants for the management of blast disease in the finger millet has shown disease reduction in the range of 16–54 % (Radjacommare et al. 2004; Kumar and Kumar 2011; Waghunde et al. 2013; Negi et al. 2015). Inoculation with the native finger millet strain MSSRFD41 resulted in 8.39 % disease incidence, which was considerably better than other treatments including a chemical fungicide. Many studies have reported that foliar application of pseudomonads has an ability to act at the site of pathogenic infestation by damaging the fungal cell wall and stopped the growth through a network of interconnecting signal resulting in accumulation of defense-related enzymes and proteins via induced systemic resistance (ISR) and systemic acquired resistance (SAR) systems (Bahadur et al. 2007; Vleesschauwer et al. 2008; Spence et al. 2014; Negi et al. 2015; Fatima and Anjum 2017; Yasmin et al. 2017).

Chitinolytic enzyme production by rhizobacteria has been attributed to antagonism against various fungal phytopathogens (Frandberg and Schnurer 1998; Vishwanathan and Samiyappan 1999). These enzymes attack on fungal cell wall and cause lysis by degrading chitin and therefore are considered as one of the most important mechanisms of biocontrol of pathogens. Fluorescent pseudomonads have also been reported to produce antifungal enzymes (Chang et al. 2003; Kohli et al. 2006; Pankaj et al. 2012). Antifungal activity of chitinase and antifungal metabolites produced by *P. fluorescens* against *P. grisea* has also been reported (Radjacommare et al. 2004; Ayyadurai et al. 2007). The suppression of ragi blast disease is done by seed treatment and foliar sprays of *P. fluorescens* even under field conditions (Patro et al. 2008; Kumar and Kumar 2011). Similarly, Karthikeyan and Gnanamanickam (2008) found that fluorescent pseudomonads could suppress 88 % of setaria blast disease under field conditions.

Blast disease was controlled by using biocontrol agents like *Trichoderma harzianum* (Gouramanis 1997) and *Pseudomonas fluorescens*. PGPR strains like *Bacillus subtilis* and *B. pumilus* have been found to control blast pathogen both via biocontrol and induction of resistance. Streptomyces species were also found to be promising for the management of blast disease (Krishnamurthy and Gnanamanickam 1998). Watanabe (1985) evaluated different *Trichoderma* sp. against 24 airborne plant pathogens including *M. oryzae* and found that isolates of *T. harzianum* and *T. viride* showed severe antagonism against *M. oryzae*, and he also found that *T. polysporum* was a weaker antagonist. The *P. fluorescens* strains showed inhibitory activity against *P. oryzae* with 47 and 59 % of mycelial inhibition (Gnanamanickam and Mew 1992). Gouramanis (1997) reported that antagonistic bioagents such as *T. harzianum* and *Chaetomium globosum* gave 70–88 % mycelial growth inhibition of *P. oryzae*. Fengycin produced by *Bacillus subtilis* was found to produce inhibitory activity against fungi *P. oryzae* (Joshi and Gardener 2006). Karthikeyan and Gnanamanickam (2008) reported effective strains of bacterial antagonists for *M. grisea* through laboratory dual-culture method. Goud and Muralikrishnan (2009) studied the antifungal activity of *P. fluorescens* against *Macrophomina phaseolina*, *Pythium ultimum*, and *P. oryzae*. All three fungi were inhibited by *P. fluorescens* with inhibitory activities ranging from 50 to 80 %. Hassanein et al. (2009) described that *Pseudomonas* sp. had the ability to produce secondary metabolites such as antibiotics, ammonia, and cyanide. Hajano et al. (2012) tested six biocontrol agents, viz. *T. harzianum, T. polysporum, T. pseudokoningii, Gliocladium virens, Paecilomyces variotii*, and *P. lilacinus*, against *M. oryzae*. Maximum mycelial inhibition was observed in *P. lilacinus* followed by *Trichoderma* spp. The efficacy of *T. viride* and *P. fluorescens* against *P. oryzae* showed growth inhibition of 72 and 78 %, respectively (Arumugam et al. 2013). *Bacillus firmus* E65 was found to be highly effective in controlling *P. oryzae* with 53.32 % while *P. aeruginosa* C32b showed 33.65 % of inhibition of the test pathogen (Suryadi et al. 2013). Pandey and Chandel (2014) studied the efficacy of *P. fluorescens* against *P. oryzae*. Maximum per cent inhibition in colony diameter was observed in *P. oryzae*. Ali and Nadarajah (2014) collected 22 *Trichoderma* isolates from soil and examined their efficacy against *M. grisea* of rice. Among the 22 isolates, 9 isolates inhibited the growth of *M. grisea* by causing 100 % coverage/overgrowth of the 9 cm plates.

3.10.4 Fungicidal Control

Ever since the first report of occurrence of blast of finger millet in India was published in 1920, a number of attempts have been made to control this disease by use of fungicides. The first authentic report of finger millet blast chemical control was by Raju and Rao (1961) who tested five fungicides and stated that Bordeaux mixture (1 %) and copper oxychloride gave best control. Subsequently, Shanmugam et al. (1962) evaluated 15 fungicides who claimed that ceresin lime dust, Dithane Z-78, flit 406 Bordeaux mixture (1 %), wettable sulfur, and zineb were used as foliar sprays. Out of the nine fungicides tested, ceresin lime mixture spray reduced blast and increased yield by 20.5 % (Vijayan and Natarajan 1967). Desh and Mohanty (1969) related the relative efficacy of different fungicides and antibiotics. Keshi and Mohanty (1970) evaluated fungicides for their efficacy to control blast and found brestonol to be effective. Benlate was found to be highly efficient in checking neck infection (Deshkar et al. 1973). Sprays with ceresin lime dust, Bla-s, zineb, and edifenphos resulted in a corresponding yield increase of 36.8, 26.3, 23.5, and 16.9 % (Sivaprakasam et al. 1974). Further on, in their continued study on control of blast of ragi they found miltox and zineb to be highly efficient (Sivaprakasam et al. 1975).

Among several chemicals, iprobenfos (IBP) was best in both controlling disease and enhancing the yield (Mohan and Jairajan 1986). With the arrival of fungicides with indirect effect like tricyclazole, the disease control was much more effective. *P. grisea* was highly sensitive to carbendazim followed by thiophanate methyl, edifenphos, kitazin, mancozeb, etc. (Kumar and Singh 1995). Many fungicides are used against blast disease, including benomyl, iprobenfos, pyroquilon, ferimzone, diclocymet, carpropamid, and metominostrobin (Kato 2001). Viswanathan and Narayanasamy (1991) reported that tricyclazole was effective at 200 mg/L in vitro against *P. oryzae* Cav. Anwar et al. (2002) observed that mancozeb exhibited excellent control of rice blast disease caused by *M. oryzae*. Mohan et al. (2011) evaluated different fungicides against *P. grisea.* Among those, tebuconazole, propiconazole, difenoconazole, tricyclazole, and azoxystrobin + difenoconazole were found significantly effective over others. Among the five fungicides, viz. thiophanate-methyl, carbendazim, fosetyl aluminum, mancozeb, and copper oxychloride, used against *M. oryzae*, only mancozeb was a highly effective fungicide that completely inhibited the mycelial growth of the pathogen at 1000 and 10,000 ppm (Hajano et al. 2012).

3.11 Conclusions and Future Prospects

From the previous conversation it is apparent that finger millet is grown in a variety of agroecological situations and it is known for resilience and drought-enduring capacity. It is relatively less prone to pests and diseases. However, climate change has impacted various biotic and abiotic constraints that limit production and productivity of small millets as well. Among the various biotic constraints, blast caused by *M. grisea* is widespread and devastating. Several researchers worked on identification of potential resistance source, effective biocontrol agents, and promising fungicides and other aspects of the disease. However, to deal with the impact of climate change on crop production, more emphasis could be given on developing cultivars tolerant to disease through marker-assisted selection (MAS), heat and salinity stress, resistance to flood and drought, modifying crop management practices, adapting new farm techniques such as resource conservation technologies, crop diversification, integrated disease management, better weather forecasting, crop insurance, and harnessing the indigenous technical knowledge of farmers which in turn results in increased production and productivity of the crop.

References

Adipala E. Seed-borne fungi of finger millet. East Afr Agric Forest J. 1992;57(3–4):173–6.

Ali H, Nadarajah K. Evaluating the efficacy of *Trichoderma* spp. and *Bacillus subtilis* as biocontrol agents against *Magnaporthe grisea* in rice. Aust J Crop Sci. 2014;8(9):1324–35.

Anonymous. Annual report of the Department of Agriculture, Uganda for the year ended 31st Dec 1958. 1959. 75 p.

Anonymous. Annual report, ICAR-AICRP on Small millets for the year 2017-18. 2018.

Anonymous. Annual report, ICAR-AICRP on Small millets for the year 2018-19. 2019.

Anwar A, Bhat GN, Singhara GN. Management of sheath blight and blast in rice through seed treatment. Ann Plant Prot Sci. 2002;10:285–7.

Arumugam K, Ramalingam P, Appu M. Isolation of *Trichoderma viride* and *Pseudomonas fluorescens* organism from soil and their treatment against rice pathogens. J Microbiol Biotechnol Res. 2013;3(6):77–81.

Ayyadurai N, Ravindra Naik P, Sakthivel N. Functional characterization of antagonistic fluorescent pseudomonads associated with rhizospheric soil of rice (*Oryza sativa* L.). J Microbiol Biotechnol. 2007;17:919–27.

Bahadur A, Singh UP, Sarma BK, Singh DP, Singh KP, Singh A. Foliar application of plant growth-promoting rhizobacteria increases antifungal compounds in pea (*Pisum sativum*) against *Erysiphe pisi*. Mycobiology. 2007;35:129–34. https://doi.org/10.4489/MYCO.2007.35.3.129.

Batsa BK, Tamang DB. Preliminary report on the study of millet diseases in Nepal. In: Maize and Finger millet, 10th Summer crops workshop, Rampur, Nepal. 1983. p. 23–8.

Bisht IS. Blast tolerance and yield loss in finger millet. Indian J Agric Sci. 1987;57(12):954–5.

Bisht IS, Bhatt JC, Chauhan PS, Joshi HC. Epidemiological studies on blast disease of ragi (Eleusine coracana) in Kumaon hills. Indian Phytopathol. 1984;37:466–8.

Bisht GS, Harish S, Singh H, Sati SC. Fungal disease of useful plants in the Garhwal Himalaya and their management. In: Saxena J, Dubey, editors. Himalayan microbial diversity part I, RC. 1997. p. 255–71.

Brondani C, Brondani RPV, Garridon LR, Ferreira ME. Development of microsatellite markers for genetic analysis of *Magnaporthe grisea*. Genet Mol Biol. 2000;23:753–62.

Burnett F. Report on agriculture in Malaya for the year 1947. 1949. p. 87.

Byregowda M, Shankaregowda BT, Seetharam A. Relationship of proteins, phenols and tannins with blast disease in finger millet (*Eleusine coracana*). Crop Improv. 1998;25:43–7.

Byregowda M, Shankaregowda BT, Seetharam A. Variability and inheritance of biochemical compounds determining resistance of blast, yield and other attributes in finger millet. Mysore J Agric Sci. 1999a;33:197–200.

Byregowda M, Seetharam A, Shankare GBT. Selection for combining grain yield with high protein and blast resistance in finger millet (*Eleusine coracana*). Indian J Genet. 1999b;59:345–9.

Cai F, Pang G, Li RX, Li R, Gu XL, Shen QR. Bio-organic fertilizer maintains a more stable soil microbiome than chemical fertilizer for monocropping. Biol Fertil Soils. 2017;53:861–72.

Chang WT, Shen CS, Wang SL. An antifungal chitinase produced by *Bacillus cereus* with shrimp and crab shell powder as a carbon source. Curr Microbiol. 2003;47:102–8.

Chaudhary RG, Vishwadhara. Epidemiology of rice blast and effect of date of sowing under upland conditions of Arunachal Pradesh. Indian Phytopathol. 1988;41:552–7.

Chauhan LS, Varma SC. Two new host records for India. Indian J Mycol Plant Pathol. 1981;11:246–7.

Dagnachew L, Villiers SDE, Masresha F, Teshome B, Tesfaye A, Geleta G, Getachew G, Kassahun T. Pathogenicity and yield loss assessment caused by *Magnaporthe oryzae* isolates in cultivated and wild relatives of finger millet (*Eleusine coracana*). Indian J Agric Sci. 2014;48(4):258–68.

Dean RA, Talbot NJ, Ebbole DJ, Farman ML, Mitchell TK, Orbach MJ. The genome sequence of rice blast fungus *Magnaporthe grisea*. Nature. 2005;434:980–6.

Desh SC, Mohanty NN. Relative efficacy of different fungicides and antibiotic along with their economics in control of blast of ragi. In: Proc. 56th Session, Indian Sci. Cong. Assn., Bombay. 1969. p. 582–3.

Deshkar MV, Sharma BL, Bhagat NK, Joshi RC. Varietal susceptibility to and evaluation of fungicides against blast of ragi. JNKVV Res J. 1973;7:295–7.

Dineshkumar SP, Seetharam A, Shankaregowda BT. Association between grain yield, grain protein content and blast disease incidence in finger millet (*Eleusine coracana*) cultivars. Milwai Newsl. 1985;4:15.

Dobinson KF, Harris RE, Hamer JE. Grasshopper, a long terminal report (LTR) retroelement in the phytopathogenic fungus, *Magnaporthe grisea*. Mol Plant-Microbe Interact. 1993;6:114–26.

Dunbar AR. The annual crop loss of Uganda. East African Literature Bureau, Dar-esalam, Nairobi, Kampala. 1969.

Fatima S, Anjum T. Identification of a potential ISR determinant from Pseudomonas aeruginosa PM12 against Fusarium wilt in tomato. Front Plant Sci. 2017;8:848. https://doi.org/10.3389/fpls.2017.00848.

Frandberg E, Schnurer J. Antifungal activity of chitinolytic bacteria isolated from air tight stored cereal grain. Can J Microbiol. 1998;44:121–7.

Gnanamanickam SS, Mew TW. Biological control of blast disease of rice (*Oryza sativa* L.) with antagonistic bacteria and its mediation by a Pseudomonas antibiotic. Ann Phytopathol Soc Jpn. 1992;58:380–5.

Goud MJP, Muralikrishnan V. Biological control of three phytopathogenic fungi by *Pseudomonas fluorescens* isolated from rhizosphere. Internet J Microbiol. 2009;7(2):223–32.
Gouramanis GD. Biological and chemical control of rice blast disease in North Greece. Cah Options Mediterraneennes. 1997;15:61–8.
Gowda SS, Gowda KTP. Epidemiology of blast disease of rice. Indian Phytopathol. 1995;38:143–5.
Gull A, Jan R, Nayik GA, Prasad K, Kumar P. Significance of finger millet in nutrition, health and value-added products: a review. J Environ Sci Comput Sci Eng Technol. 2014;3:1601–8.
Gupta U. Effect of dates of sowing in the incidence and severity of blast disease in finger millet. In: Extended summary, National Seminar on Sm. all Millets, TNAU, 23–23 April. 1997. p. 71.
Gupta PK, Varshney RK. The development and use of microsatellite markers for genetic analysis and plant breeding with emphasis on bread wheat. Euphytica. 2000;113:163–85.
Gupta R, Bisaria VS, Sharma S. Effect of agricultural amendments on *Cajanus cajan* (Pigeon pea) and its rhizospheric microbial communities—a comparison between chemical fertilizers and bioinoculants. PLoS One. 2015;10:e0132770.
Hajano J, Lodhi AM, Mumtaz AP, Khanzada MA, Shah SG. *In vitro* evaluation of fungicides, plant extracts and biocontrol agents against rice blast pathogen *Magnaporthe oryzae* Couch. Pak J Bot. 2012;44(5):1775–8.
Hassanein WA, Awny NM, El-Mougith AA, El-Dien SH. The antagonistic activities of some metabolites produced by *Pseudomonas aeruginosa* sha8. J Appl Sci Res. 2009;5:404–14.
Hawkins NJ, Cools HJ, Sierotzki H, Shaw MW, Knogge W, Kelly SL. Paralog re-emergence: a novel, historically contingent mechanism in the evolution of antimicrobial resistance. Mol Biol Evol. 2014;31:1793–802.
Hebert TT. The perfect stage of *Pyricularia grisea*. Phytopathology. 1971;61:83–7.
Hilu K, De Wet J, Harlan J. Archaeobotanical studies of *Eleusine coracana* ssp. *coracana* (finger millet). Am J Bot. 1979;66(3):330–3.
Hollomon DW. Fungicide resistance: facing the challenge; a review. Plant Prot Sci. 2016;51:170–6.
Inukai T, Vales MI, Hori K, Sato K, Hayes PM. *RM1* confers blast resistance in barley and is located within the complex of resistance genes containing *Mla*, a powdery mildew resistance gene. Mol Plant-Microbe Interact. 2006;19:1034–41.
Jain AK, Yadava HS. Mechanism of blast resistance in finger millet (*Eleusine coracana* l. gaertn.). Trop Agric. 2004;81(2):73–9.
Jegan S, Kathiravan R, Purushothaman D, Vaiyapuri PR. Potential of finger millet indigenous rhizobacterium Pseudomonas sp. MSSRFD41 in blast disease management—growth promotion and compatibility with the resident rhizomicrobiome. Front Microbiol. 2018;9:1029.
Joshi R, Gardener BM. Identification and characterization of novel genetic markers associated with biological control activities in *Bacillus subtilis*. Phytopathology. 2006;96(2):145–54.
Karthikeyan V, Gnanamanickam SS. Biological control of Setaria blast (*M. grisea*) with bacterial strains. Crop Prot. 2008;27(2):263–7.
Kato H. Rice blast control. Pesticide Outlook 23-25. Japan crop protection association (JCPA), Japan. 2001.
Kato H, Yamaguchi T, Nishihara N. Seed transmission pathogenicity and control of ragi blast fungus and susceptibility of ragi to *Pyricularia* spp. from grasses, cereals and mioga. Ann Phytopathol Soc Jpn. 1977;43:392–401.
Kaye C, Milazzo J, Rozenfeld S, Lebrun MH, Tharreau D. The development of simple sequence repeat (SSR) markers for *Magnaporthe grisea* and their integration into an established genetic linkage map. Fungal Genet Biol. 2003;40:207–14.
Keshi KC, Mohanty NN. Efficacy of different fungicides and antibiotic on the control of blast of ragi. In: Plant Dis Problems Proc I Intr Symp Plant Path IPS New Delhi. 1970. p. 475–34.
Kim NS, Park NI, Kim SU, Kim ST, Han SS, Kang KY. Isolation of TC/AG repeat microsatellite sequences for fingerprinting rice blast fungus and their possible horizontal transfer to plant species. Mol Cells. 2000;10:127–34.
Kohli U, Dogra RC, Kapoor KK, Sharma PK. Chitinolytic *Pseudomonas maltophilia* antagonistic to root rot pathogens of sunflower (*Helianthus annus*). Indian J Microbiol. 2006;46:45–9.

Krishnamurthy K, Gnanamanickam SS. Biological control of rice blast by *Pseudomonas fluorescens* strain Pf7-14: evaluation of a marker gene and formulations. Biol Control. 1998;3:158–65.

Krishnappa P, Ramaswamy GR, Jyothi V. Role of silica fertilizers on blast (*Pyricularia grisea* Sacc.) disease development in finger millet [*Eleusine coracana* (L.) Gaertn.]. Plant Dis Res. 2013;28(2):207–8.

Kumar B, Kumar J. Management of blast disease of finger millet (*Eleusine coracana*) through fungicides, bioagents and varietal mixture. Indian Phytopathol. 2011;64:272.

Kumar A, Singh RA. Biochemical studies among different isolates of *Magnaporthe grisea* from different hosts. Indian Phytopathol. 1995;48:132–6.

Kumar BS, Kumar TBA, Nagaraju. Epidemiological studies of neck and finger blast disease in finger millet (*Eleusine coracana* (L) Gaertn.) caused by *Pyricularia grisea* (Cke) Sacc. Environ Ecol. 2005;23S(Special 4):861–3.

Kumar VBS, Amruta BS, Nagaraju. Virulence analysis of *Pyricularia grisea* (Cke.) Sac. on different finger millet genotypes and cultivars to determine race differential lines. Environ Ecol. 2007;25(1):190–2.

Kuwite CA, Shao FM. *Pyricularia* spp. causing head blight of finger millet (*Eleusine coracana*) and other fungi associated with finger millet in Tanzania. In: de Millians WAJ, Frederiksen RA, Bengston GD, editors. Sorghum and millets diseases: a second world review. Patancheru: ICRISAT; 1992.

Latha AM, Rao KV, Reddy VD. Production of transgenic plants resistant to leaf blast disease in finger millet (*Eleusine coracana* (L.) Gaertn.). Plant Sci. 2005;169(4):657–67.

Levy M, Fernando J, Correa-Victoria, Zeigler RS, Xu S, Hamer JE. Genetic diversity of the rice blast fungus in a disease nursery in Colombia. Phytopathology. 1993;83:1427–33.

Li CY, Li JB, Liu L, Yang J, Su Y, Wang YY, Xie Y, Ye M, Zhu YY. Development of minisatellite markers in phytopathogenic fungus, *Magnaporthe grisea*. Mol Ecol Notes. 2007;7(6):978–80.

Madhukeshwara SS, Mantur SG, Ramanathan A, Kumar J, Shashidhar VR, Jagadish PS, Seenappa K, Anilkumar TB. On-farm adaptive management of the blast of finger millet. Int Sorghum Millets Newsl. 2005;46:111–4.

Magar PB, Acharya B, Pandey B. Use of chemical fungicides for the management of rice blast (*Pyricularia grisea*) disease at Jyotinagar, Chitwan, Nepal. Int J Appl Sci Biotechnol. 2015;3:5–10.

Mc Rae W. Detailed administration report of the government mycologist for the year 1919–20. 1920. Madras Agri. Dept.

McRae W. Report of the imperial mycologist. Agric. Res. Inst. Pusa Scientific Reports 1921–22, 44-50 psmioga. Ann Phytopathol Soc Jpn. 1922;43:392–401.

Mishra SK, Patnaik HB, Jena BC. Influence of cultivation methods on the incidence of blast disease in ragi. Milwai Newsl. 1985;4:17.

Mohamed MI. A disease of *Eleusine coracana* new to Somalia. Rev Agric Sutro Trop. 1980;74:73–9.

Mohan L, Jairajan R. Chemical control of ragi blast. Madras Agric J. 1986;73:709–10.

Mohan C, Amrinder K, Sandeep R. *In vitro* evaluation of different fungicides against *Pyricularia grisea*. Plant Dis Res. 2011;26(2):178.

Murakami J, Tosa Y, Kataoka T, Tomita R, Kawasaki J, Chuma I, Sesumi Y, Kusaba M, Nakayashiki H, Mayama S. Analysis of host species specificity of *Magnaporthe grisea* toward wheat using a genetic cross between isolates from wheat and foxtail millet. Phytopathology. 2000;90:1060–7.

Muyanga S, Danial DL. Production and research review of small millets cereals in Zambia. In: Daniel DL, editor. Breeding for disease resistance with emphasizing durability. Netherlands: Wageningen Agril. Univ. Wageningen; 1995. p. 60–4.

Nagaraja A, Jagadish PS, Ashok EG, Gowda KTK. Avoidance of finger millet blast by ideal sowing time and assessment of varietal performance under rain fed production situations in Karnataka. J Mycopathol Res. 2007;45(2):237–40.

Nagaraja A, Reddy Y, Gowda J, Reddy BA. Association of plant characters and weather parameters with linger millet blast. Crop Res. 2010;39(1/2/3):123–6.

Negi YK, Prabha D, Garg SK, Kumar J. Biological control of ragi blast disease by chitinase producing fluorescent Pseudomonas isolates. Org Agric. 2015;7:63–71. https://doi.org/10.1007/s13165-015-0142-2.

Ou SH. Pathogen variability and host resistance in rice blast disease control. Annu Rev Phytopathol. 1980;18:167–87.

Ou SH, Ayad MR. Pathogenic races of *Pyricularia oryzae* derived from monoconidial cultures. Phytopathology. 1968;58:179–82.

Ou SH, Nuque FL, Ebron TT, Awodens VA. Pathogenic races of *Pyricularia oryzae* derived from monoconidial cultures. Plant Dis Rep. 1970;54:1045–9.

Pall BS. Assessment of losses due to neck blast of ragi (*Eleusine coracana* (L) Gaertn). Food Farm Agric. 1977;9:55.

Pall BS. Effect of seed borne inoculum of *Pyricularia setariae* Nishikado on the finger millet blast. Agric Sci Dig. 1988;8:225–6.

Pall BS. Biochemical studies on blast disease of finger millet (*Eleusine coracana*). Bioved. 1992;3:153–4.

Pandey SK, Chandel SCR. Efficacy of *Pseudomonas* as biocontrol agent against plant pathogenic fungi. Int J Curr Microbiol App Sci. 2014;3(11):493–500.

Pankaj K, Maheshwari DK, Dubey RC. *Bacillus* strains isolated from rhizosphere showed plant growth promoting and antagonistic activity against phytopathogens. Microbiol Res. 2012;167:493–9.

Park M. Report of the work done of the mycological division. Admn Rep Dir Agric Ceylon. 1932;21:103–11.

Patel RP, Tripathi SK. Epidemiology of blast of finger-millet caused by *Pyricularia grisea* (cke) Sacc. Adv Plant Sci. 1998;11(1):73–5.

Patro TSSK, Rani CH, Kumar GV. *Pseudomonas fluorescens*, a potential bioagent for the management of blast in *Eleusine coracana*. J Mycol Plant Pathol. 2008;38(2):298–300.

Prajapati VP, Sabalpara AN, Pawar DM. Assessment of yield loss due to finger millet blast caused by *Pyricularia grisea* (Cooke) Sacc. Biosci Trends. 2013;6(6):876–8.

Radjacommare R, Ramanathan A, Kandan A, Harish S, Thambidurai G, Sible GV, Ragupathi N, Samiyappan R. PGPR mediates induction of pathogenesis-related (PR) proteins against the infection of blast pathogen in resistant and susceptible ragi [*Eleusine coracana* (L.) Gaertner] cultivars. Plant Soil. 2004;266(1/2):165–76.

Raju DG, Rao PG. Field testing of fungicides. Andhra Agric J. 1961;8:130–4.

Ramakrishnan KV. Studies on the morphology, physiology and parasitism of the genus *Pyricularia* in Madras. Proc Indian Acad Sci Scit B. 1948;27:174–93.

Ramappa HK, Ravishankar CR, Prakash P. Estimation of yield loss and management of blast disease in finger millet (Ragi). In: Proceedings of Asian Conference Mycology Plant Pathology, University of Mysore. 2002. p. 195.

Ramappa HK, Ravishankar CR, Prakash P, Revanna HP. Effect of sowing time on the leaf blast severity of finger millet in nursery. Environ Ecol. 2006;24S(Special 1):95–6.

Rao ANS. Estimation of crop loss in finger millet *(Eleusine coracana* (L) Gaertn.) due to blast disease (*Pyricularia grisea* (Cke) Sacc). Ph.D. thesis, UAS, Bangalore. 1981. p. 113.

Rao ANS. Estimates of losses in finger millet (*Eleusine coracana*) due to blast disease (*Pyricularia grisea*). Mysore J Agric Sci. 1990;24:57–60.

Rao ANS, Hegde RK. Effect of disease incidence on the grain yield of ragi due to blast. Curr Res. 1987;16:26–7.

Rath GC, Mishra D. Nature of losses due to neck blast infection in ragi. Sci Cult. 1975;41:322.

Rathour R, Singh BM, Sharma TR. Sexual fertility in *Magnaporthe grisea* isolates from rice, finger millet and crabgrass in Himachal Pradesh. J Mycol Plant Pathol. 2004;34:230–2.

Rathour R, Sharma R, Sharma V. Genetic differentiation of rice and non-rice populations of *Magnaporthe grisea* from North-Western Himalayas using native protein and isozyme polymorphisms. J Phytopathol. 2006;154:641–7.

Ravikumar RL, Seetharam A. Character association in segregating population of finger millet (*Eleusine coracana*) in blast–endemic region. Indian J Agric Sci. 1993;63:96–9.

Roy AK. Further records of plant disease from Karbi Anglong dist. Assam J Res. 1989;10:88–91.

Santhanam R, Luu VT, Weinhold A, Goldberg J, Oh Y, Baldwin IT. Native root-associated bacteria rescue a plant from a sudden-wilt disease that emerged during continuous cropping. Proc Natl Acad Sci U S A. 2015;112:E5013–20. https://doi.org/10.1073/pnas.1505765112.

Schreiter S, Ding GC, Grosch R, Kropf S, Antweiler K, Smalla K. Soil type-dependent effects of a potential biocontrol inoculant on indigenous bacterial communities in the rhizosphere of field-grown lettuce. FEMS Microbiol Ecol. 2014;90:718–30. https://doi.org/10.1111/1574-6941.12430.

Sekar J, Raj R, Prabavathy VR. Microbial consortial products for sustainable agriculture: commercialization and regulatory issues in India. In: Singh HB, Sarma BK, Keswani C, editors. Agriculturally important microorganisms. Singapore: Springer; 2016. p. 107–32.

Selvaraj S, Ganeshamoorthi P, Anand T, Raguchander T, Seenivasan N, Samiyappan R. Evaluation of a liquid formulation of *Pseudomonas fluorescens* against *Fusarium oxysporum* f. sp. *cubense* and *Helicotylenchus multicinctus* in banana plantation. Biocontrol. 2014;59:345–55. https://doi.org/10.1007/s10526-014-9569-8.

Shanmugam N, Mariappa V, Kandaswamy M. Ragi blast and its control. Madras Agric J. 1962;49:77–80.

Sharma D, Jamra G, Singh UM, Sood S, Kumar A. Calcium biofortification: three pronged molecular approaches for dissecting complex trait of calcium nutrition in finger millet (*Eleusine coracana*) for devising strategies of enrichment of food crops. Front Plant Sci. 2017;7:2028. https://doi.org/10.3389/fpls.2016.02028.

Sivaprakasam K, Pillayarsami K, Iyemperumal S, Jaleel AN. Note on the effect of fungicides on the control of finger millet blast. Indian J Agric Sci. 1974;44:249–50.

Sivaprakasam K, Pillayarsami K, Lyemperumal S, Ahmed N. Efficacy of some chemicals in the control of finger millet blast. Madras Agric J. 1975;62:84–6.

Somappa KM. Mechanism of resistance and biocontrol of blast of ragi caused by *Pyricularia grisea* (Cke) Sacc. M.Sc. (Agri.) Thesis UAS, Bangalore. 1999.

Sonah H, Deshmukh RK, Parida SK, Chand S, Kotasthane A. Morphological and genetic variation among different isolates of *Magnaporthe grisea* collected from Chhattisgarh. Indian Phytopathol. 2009;62(4):469–77.

Soubabere O, Jorge V, Nottendheen JVL. Sequence characterized amplified region markers for the rice blast fungus *Magnaporthe grisea*. Mol Ecol Notes. 2001;1:19–21.

Spence C, Alff E, Johnson C, Ramos C, Donofrio N, Sundaresan V. Natural rice rhizospheric microbes suppress rice blast infections. BMC Plant Biol. 2014;14:130. https://doi.org/10.1186/1471-2229-14-130.

Sundaram NV, Palmer CT, Nagarajan K, Presscot JM. Disease survey of sorghum and millet in India. Plant Dis Rep. 1972;56:740–3.

Suryadi Y, Susilowati DN, Riana E, Mubarik NR. Management of rice blast disease using formulated bacterial consortium. Emir J Food Agric. 2013;25(5):349–57.

Suzuki F, Suga H, Tomimura K, Fuji S, Arai M, Koba A, Nakajima T. Development of simple sequence repeats markers for Japanese isolates of *Magnaporthe grisea*. Mol Ecol Resour. 2009;9:588–90.

Takan JP, Akello B, Esele P, Manyasa EO, Obilana AB, Audi PO, Kibuka J, Odendo M, Oduori CA, Ajanga S, Bandyopadhyay R, Muthumeenakshi S, Coll R, Brown AE, Talbot NJ, Sreenivasaprasad S. Finger millet blast pathogen diversity and management in East Africa: a summary of project activities and outputs. Int Sorghum Millets Newsl. 2004;45:66–9.

Takan JP, Chipili J, Muthumeenakshi S, Talbot NJ, Manyasa EO, Bandyopadhyay R, Sere Y, Nutsugah SK, Talhinhas P, Hossain M, Brown AE, Sreenivasaprasad S. *Magnaporthe oryzae* populations adapted to finger millet and rice exhibit distinctive patterns of genetic diversity, sexuality and host interaction. Mol Biotechnol. 2011;50:145. https://doi.org/10.1007/s12033-011-9429-z.

Tanaka M, Nakayashiki H, Tosa Y. Population structure of *Eleusine* isolates of *Pyricularia oryzae* and its evolutionary implications. J Genet Plant Pathol. 2009;75:173–80.

Thakur RP, Sharma R, Rai KN, Gupta SK, Rao VP. Screening techniques and resistance sources for foliar blast in pearl millet. J SAT Agric Res. 2009;7:5.

Thirumalachar MJ, Mishra JN. Contribution to the study of fungi of Bihar, India—I. Sydowia. 1953;7(14):29–83.

Thomas KM. Detailed administration report of the govt. Mycologist Madras for the year 1939-40. 1940. 18 p.

Thomas P, Sekhar AC. Effects due to rhizospheric soil application of an antagonistic bacterial endophyte on native bacterial community and its survival in soil: a case study with *Pseudomonas aeruginosa* from banana. Front Microbiol. 2016;7:493.

Thompson A. Notes on plant diseases in 1940. Malay Agric J. 1941;24:241–5.

Upadhyaya HD, Pundir RPS, Gowda CLL. Genetic resources diversity of finger millet—a global perspective. In: Mgonja MA, Lenne JM, Manyasa E, Sreenivasaprasad S, editors. Finger millet blast management in East Africa. Creating opportunities for improving production and utilization of finger millet. Patancheru: International Crops Research Institute for the Semi-Arid Tropics; 2007. p. 90. ISBN: 978-92-9066-505-2.

Varshney RK, Chabane K, Hendre PS, Agarwal R, Graner A. Comparative assessment of EST-SSR, EST-SNP and AFLP markers for evaluation of genetic diversity and conservation of genetic resources using wild, cultivated and elite barleys. Plant Sci. 2007;173:638–49.

Venkatarayan SV. Report of the work done in the mycological section during the year 1935-36 Admn Rep Agric Dept Mysore (1935-36). 1937. p. 51–5.

Venkatarayan SV. Diseases of ragi. Mysore J Agric Sci. 1946;24:50–7.

Vijayan KM, Natarajan S. A note on the fungicidal control of ragi blast at Tindivanam. Madras Agric J. 1967;54:485–6.

Viji G, Gnanamanikam SS. Mating type distribution and futility status of *Magnaporthe grisea* populations from various hosts in India. Plant Dis. 1998;82(1):36–40.

Viji G, Gnanamanickam SS, Levy M. DNA polymorphisms of isolates of *Magnaporthe grisea* from India that are pathogenic to finger millet and rice. Mycol Res. 2000;104(2):161–7.

Vishwanath S, Sanne GS, Seetharam A, Shankare GBT. Reaction to blast disease of released and pre-released varieties of finger millet from different states. Millet News Lett. 1986;5:31.

Vishwanath S, Mantur SG, Channamma KAL. Recent approaches in the management of finger millet diseases. In: Proc. National Seminar on Small Millets. ICAR and TNAU, Coimbatore. 1997. p. 27–30.

Vishwanathan R, Samiyappan R. Induction of systemic resistance by plant growth promoting rhizobacteria against red rot diseases caused by *Colletotrichum falcatum* went in sugarcane. Proc Sugar Technol Assoc India. 1999;61:24–39.

Viswanathan R, Narayanasamy P. Effect of tricyclazole on the physiology of *Pyricularia oryzae*. Z Pflanzenkrankh Pflanzenschutz (Germany, F.R.). 1991;98(2):205–12.

Vleesschauwer D, Djavaheri M, Bakker PA, Hofte M. *Pseudomonas fluorescens* WCS374r-induced systemic resistance in rice against *Magnaporthe oryzae* is based on pseudobactin-mediated priming for a salicylic acid-repressible multifaceted defense response. Plant Physiol. 2008;148:1996–2012. https://doi.org/10.1104/pp.108.127878.

Vos P, Hogers R, Bleeker M, Reijans M, Lee T, Hornes M, Friters A, Pot J, Paleman J, Kuiper M, Zabeau M. AFLP: a new technique for DNA fingerprinting. Nucleic Acids Res. 1995;23:4407–14.

Waghunde RR, Sabalpara AN, Deshmukh AJ, Pandya JR. Characterization and evaluation of native *Pseudomonas* spp. of south Gujarat against leaf blast in finger millet. J Biol Control. 2013;27(4):312–8.

Wallace GB, Wallace M. Tanganyika territory fungus list. Recent research, XI Mycol Cir Dep Agric Tanganyika. Vol. 5. 1948. 24 p.

Wang X, Mavrodi DV, Ke L, Mavrodi OV, Yang M, Thomashow LS. Biocontrol and plant growth-promoting activity of rhizobacteria from Chinese fields with contaminated soils. Microb Biotechnol. 2015;8:404–18. https://doi.org/10.1111/1751-7915.12158.

Watanabe N. Antagonism by various kinds of *Trichoderma* fungi to airborne plant pathogens. Bull Fac Agric Meiji Univ. 1985;68:1–9.

Welsh J, McClelland M. Fingerprinting genomes using PCR with arbitrary primers. Nucleic Acids Res. 1990;18:7213–8.

Williams JGK, Kubelik AR, Livak KJ, Rafalski JA, Tingey SV. DNA polymorphisms amplified by arbitrary primers are useful as genetic markers. Nucleic Acids Res. 1990;18:6531–5.

Yaegashi H, Nishihara N. Production of the perfect stage in *Pyricularia* from cereals and grasses. Ann Phytopathol Soc Jpn. 1976;42:511–5.

Yaegashi H, Udagawa S. The taxonomical identity of the perfect stage of *Pyricularia grisea* and its allies. Can J Bot. 1978;56:180–3.

Yasmin S, Hafeez FY, Mirza MS, Rasul M, Arshad HMI, Zubair M. Biocontrol of bacterial leaf blight of rice and profiling of secondary metabolites produced by rhizospheric *Pseudomonas aeruginosa* BRp3. Front Microbiol. 2017;8:1895. https://doi.org/10.3389/fmicb.2017.01895.

Yin D, Wang N, Xia F, Li Q, Wang W. Impact of biocontrol agents *Pseudomonas fluorescens* 2P24 and CPF10 on the bacterial community in the cucumber rhizosphere. Eur J Soil Sci. 2013;59:36–42. https://doi.org/10.1016/j.ejsobi.2013.09.001.

Zheng Y, Zhang G, Lin F, Wang Z, Jin G, Yang L, Wang Y, Chen X, Xu Z, Zhao X, Wang H, Lu J, Lu G, Wu W. Development of microsatellite markers and construction of genetic map in rice blast pathogen *Magnaporthe grisea*. Fungal Genet Biol. 2008;45(10):1340–7.

Chapter 4
Finger Millet Blast Disease: Potential Threat to Global Nutrition Security

B. Jeevan, Rajashekara Hosahatti, K. K. Mishra, A. R. N. S. Subbanna, Ashish Kumar Singh, and Devender Sharma

Contents

4.1 Origin, Distribution and Diversity

Finger millet blast disease is caused by the haploid, filamentous, ascomyceteous fungus *M. grisea* (anamorph *Pyricularia grisea*). Blast disease has emerged as an explosive threat to many of the landraces and high-yielding varieties of finger millet and can be able to cause more than 80% yield losses under congenial environmental conditions (Vishwanath et al. 1986). The disease has been prevalent in semiarid regions of Africa including Kenya, Uganda, Tanzania, Rwanda, Ethiopia, Zambia, and Nigeria. In Asia the disease has been reported from India, China, Nepal, Sri

B. Jeevan (✉) · R. Hosahatti · K. K. Mishra · A. R. N. S. Subbanna · A. K. Singh · D. Sharma
ICAR-Vivekananda Parvatiya Krishi Anusandhan Sansthan, Almora, Uttarakhand, India
e-mail: jeevan.b@icar.gov.in

S. C. Nayaka et al. (eds.), *Blast Disease of Cereal Crops*, Fungal Biology,
https://doi.org/10.1007/978-3-030-60585-8_4

Lanka, and Saudi Arabia. *Pyricularia grisea* was first isolated from crabgrass (*Digitaria sanguinalis*) almost two centuries ago (Saccardo 1880). In India, finger millet blast was reported for the first time from Tanjore delta of Tamil Nadu (McRae 1920). During 1982, the pathogen was also isolated from rice and named as *Pyricularia oryzae* (Cavara, Fungi Longobardiae #49). Despite few morphological dissimilarities between them, these scanty variations were not considered sufficient to differentiate them. However, *M. grisea* was considered to use as a name to represent the *Mg.* complex as per the rules of nomenclature.

The pathogen can infect more than 50 host species in the family Poaceae, including rice, wheat, pearl millet, foxtail millet, and finger millet (Ou 1985; Rossman et al. 1990). In spite of having a wide host range, pathogen populations exist to adapt to their specific hosts and are capable of infecting a single host (Todman et al. 1994; Viji et al. 2000). However, some researchers have reported that few isolates were successful in cross infection under experimental conditions (Mackill and Bonman 1986; Kumar and Singh 1995) while others failed to confirm the results (Todman et al. 1994). The genus *Magnaporthe* comprises five different species, viz. *M. grisea*, *M. oryzae*, *M. poae*, *M. rhizophila*, and *M. salvinii*, which shared common morphological features such as three-septate spindle-shaped ascospores and black ascoma with long hairy necks. Nevertheless, the phylogenetic analyses using molecular sequence data of actin, calmodulin, and beta-tubulin genes resolved the isolates from crabgrass as well-defined phylogenetic group from rice and other grass isolates. The crabgrass isolates were named as *M. grisea* and the isolates from rice and other grasses were described as *M. oryzae* (Couch and Kohn 2002).

4.2 Taxonomy and Biology of Pathogen

Magnaporthe grisea belongs to the family Magnaporthaceae and is an ascomyceteous fungus because it produces ascospores in a sexual spore-bearing cell called asci. The asci are produced within the specialised fruiting structures known as perithecia (Fig. 4.1). The fungi are haploid; mycelium is septate having nuclei within the mycelium.

4.2.1 Taxonomic Position

Kingdom: Fungi
Phylum: Ascomycota
Class: Sordariomycetes
Subclass: Sordariomycetidae
Family: *Magnaporthaceae*
Genus: *Magnaporthe*
Species: *M. grisea*

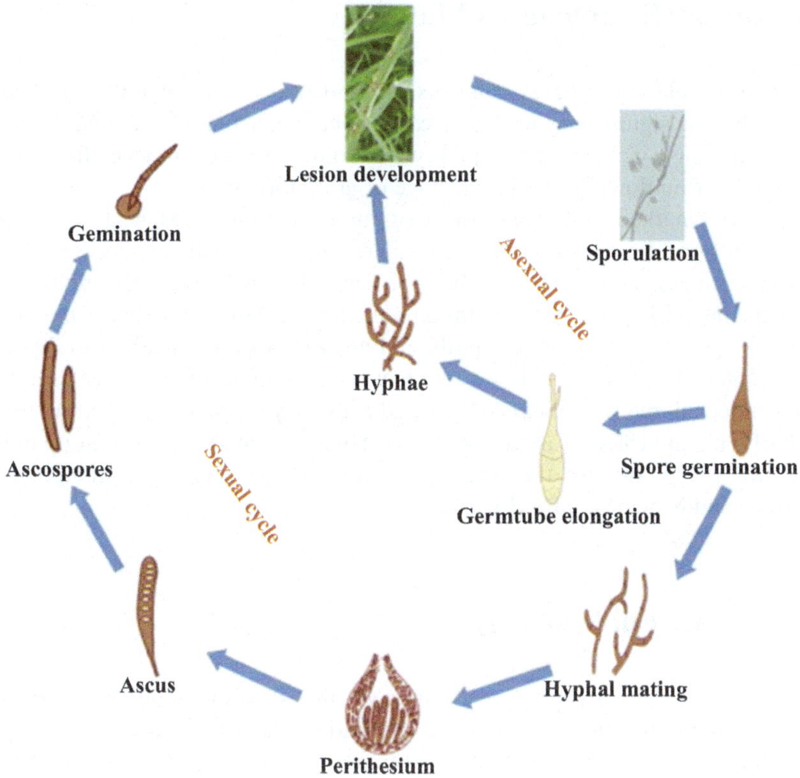

Fig. 4.1 Life cycle of *Magnaporthe grisea*

4.2.2 Sexual Reproduction

The teleomorphic stage of the fungi is generally produced if opposite mating type is paired, but the rate of occurrence of sexual reproduction is uncommon. The fungus produces spindle-shaped sexual spores with three septa. The asci are unitunicate and the fungus is heterothallic with a bipolar mating system (single-mating-type locus exists with two alleles). However, the alleles include genes encoding for completely distinct proteins; hence, they are technically not alleles, and are described as idiomorphs (Glass et al. 1990).

4.2.3 Asexual Reproduction

The conidia of the pathogen are pyriform, three celled, and hyaline and are produced on the top of conidiophores. The conidia germinate, producing a thin germ tube, which elongates and differentiates into an appressorium. A slender penetration peg forms at the base, which enters the cuticle and manifests within the host tissue.

4.3 Disease Symptoms and Losses

Finger millet blast is a devastating disease and impacts finger millet production drastically. The significance of this disease is obtained from the fact that the pathogen can minimise yield and grain quality. Maximum losses occur when finger infection starts during flowering or initial stage of grain formation.

Disease is marked by the appearance of tiny lesions on leaves, neck, and fingers. On leaves, typical eyespot-shaped lesions are formed, which are broadened in the centre and tapered at both ends of the spot. These lesions are usually greyish in the centre having dark brown margin. Initially the leaves show chlorosis; as the disease advances these lesions enlarge rapidly and coalesce together leading to drying of leaves. The pathogen is also known to infect neck region resulting in neck rot. If the neck is infected, the above parts of infected neck may become dry resulting in total death of the plant (Sreenivasaprasad 2004). This may cause yield losses up to 90% (Ekwamu 1991). In case the fingers are infected, the seeds get shrivelled and deformed and become chaffy (Fig. 4.2).

4.4 Disease Epidemiology

Blast disease severity greatly depends upon weather situations, cultivar and infected plant part. A combination of cloudy weather and frequent drizzling, which supports leaf wetness for a longer time with an optimum temperature of 25–28 °C, favours disease development. The presence of blast spores in the air throughout the year, especially in the tropical conditions, favours the occurrence of disease. The pathogen can also survive in soils and establishes better in the soils having high N_2 content (Sreenivasaprasad 2004; Hayden 1999).

The primary source of inoculum includes grasses, infected plant debris and infested seeds on the soil. The leftover infested seeds on the soil can produce spores abundantly during the initial stage of the crop which are easily disseminated and deposited on healthy leaves especially under windy conditions. These spores germinate and invade the leaf tissues. Disease extremity is often correlated with the

Fig. 4.2 Disease symptoms of finger millet blast: (**a**) leaf blast, (**b**) neck blast, (**c**) finger blast

amount of primary inoculum available. The amount of disease at the reproductive stage is influenced by the quantum of disease at the end of vegetative phase of the crop. The secondary spores produced at the end of vegetative phase may infect the neck and cause neck blast as the disease advances; the pathogen may also infect the panicles, leading to finger blast. Panicle blast phase of the disease is considered as the most destructive stage, since the infection at this stage can affect the entire panicle resulting in poor seed setting and seeds can also be infected.

4.5 Genetics of Disease Resistance

Genetic resistance is the best possible way to combat the disease because the crop is largely cultivated by subsistence farmers who cannot afford the disease management through expensive chemicals and fungicides that have shown limited efficacy. Disease resistance is generally governed by specific interaction between resistance (R) gene in the host and corresponding gene which conditions avirulence (*Avr*) in the pathogen. Approximately, 48 R-genes have been cloned from various plant species including rice, wheat, etc. Among them, a large number of R-genes share maximum similarities with nucleotide-binding site and leucine-rich repeat (NBS–LRR) domain protein sequences.

There is very limited information available on genetics of resistance to finger millet blast disease. However, efforts are being made to identify the markers linked to R-genes/QTL conferring resistance to blast disease in finger millet. Panwar et al. (2011) identified NBS-09711, NBS-07688, NBS-05504, NBS-03509 and EST-SSR-04241 markers which are potentially related to blast resistance gene from resistant finger millet genotypes. Studies have also reported that the genotypes VHC3997, VHC3996 and VHC3930 were found to be highly resistant. The molecular markers linked to R-genes/QTLs can be used for cloning of complete gene, which can be used in the marker-assisted breeding for introgression of the blast resistance alleles in finger millet breeding programmes (Babu et al. 2014).

4.6 Disease Management

4.6.1 Management Through Cultural Approaches

Sanitation of field by removing infected straw and crop debris will not only help to reduce the disease propagules but also avoid the spread of disease. Use of disease-free or certified seeds will reduce the source of primary inoculum which helps to decrease the disease level. Planting time is more important because sowing seeds with the onset of rainy season significantly reduces the early infection of seedlings. Avoid excessive use of fertiliser as it increases the disease. Maintaining optimum plant density is highly recommended, since high-density planting may increase disease development. Weed management plays a crucial role in disease management, since presence of weeds deteriorates crop growth which helps the pathogen for easy infection and frequent weeding may also eliminate alternate hosts of blast pathogen. Intercropping of finger millet with cowpea, groundnut and pigeon pea effectively reduces the disease. Planting of improved varieties such as GPU 28, IE 2911, IE 2957, VHC 3997, VHC 3996 and VHC 3930 with good agronomic practices will significantly reduce the blast disease.

4.6.2 Management Through Bioagents

Biological control is an important component in plant disease management which involves suppression of plant pathogens using beneficial microorganisms without harming the environment. These bioagents exhibit a number of mechanisms such as antibiosis, induced resistance, competition, production of lytic enzymes, HCN and siderophore, which not only suppresses pathogens but also promotes better plant growth. Application of *Pseudomonas fluorescens* isolate Pf-30 showed more than 80% inhibition of *Magnaporthe grisea* (Negi et al. 2015). Application of *P. fluorescens* and *Trichoderma harzianum* in combination resulted in significant reduction of blast disease incidence caused by *Magnaporthe grisea* (Netam et al. 2016). The use of bioagents has been considered as an alternative to chemical fungicides; it aims to reduce the dependence on plant protection chemicals and their hazardous effects on ecosystem.

4.6.3 Management Through Fungicides

Seed treatment with tricyclazole at 1 g/kg of seeds will inhibit spore germination and mycelial growth and reduce the spread of disease. Systemic fungicides such as pyroquilon and tricyclazole were found to be the most ideal and effective chemicals to reduce both leaf blast and neck blast under field conditions.

4.7 Conclusion and Future Prospects

Blast disease remains a threat to the production of finger millet worldwide, due to the variability of the pathogen and its capability to overcome the host resistance. The biggest challenge for the finger millet blast disease is lack of identified R-genes and inefficiency of the fungicides when the disease pressure is too high with favourable weather conditions for the development of the disease. This requires proper phenotyping facilities, identification of resistance sources and integrated disease management approaches to reduce yield losses. Besides, there is a need for collaborative research and exchange of germplasm resources throughout the globe to combat this disease.

References

Kalyana Babu B, Dinesh P, Agrawal PK, Sood S, Chandrashekara C, Bhatt JC, Kumar A, Prasad M. Comparative genomics and association mapping approaches for blast resistant genes in finger millet using SSRs. PLoS One. 2014;9(6):e99182.

Couch BC, Kohn LM. A multilocus gene genealogy concordant with host preference indicates segregation of a new species, *Magnaporthe oryzae*, from *M. grisea*. Mycologia. 2002;94:683–93.

Ekwamu A. Influence of head blast infection on seed germination and yield components of finger millet (*Eleusine coracana* L. Gaertn.). Trop Pest Manag. 1991;37:122–3.

Glass NL, Grotelueschen J, Metzenberg RL. *Neurospora crassa* A mating-type region. Proc Natl Acad Sci U S A. 1990;87:4912–6. https://doi.org/10.1073/pnas.87.13.4912.

Hayden NJ. An investigation into the biology, epidemiology and management of finger millet blast in low-input farming systems in East Africa. DFID Crop Protection Programme, Final Technical Report, Project R6733. Chatham, Kent, UK: Natural Resources Institute (NRI), University of Greenwich. 1999; 81 pp.

Kumar A, Singh RA. Differential response of *Pyricularia grisea* isolates from rice, finger millet and pearl millet to media, temperature, pH and light. J Mycol Plant Pathol. 1995;25(3):238e42.

Mackill AO, Bonman JM. New hosts of *Pyricularia oryzae*. Plant Dis. 1986;70:125e7.

McRae W. Detailed administration report of the government mycologist for the year 1919–20. 1920. Madras Agri Dept

Negi YK, Prabha D, Garg SK, Kumar J. Biological control of ragi blast disease by chitinase producing fluorescent *Pseudomonas* isolates. Org Agric. 2015;7:63–71. https://doi.org/10.1007/s13165-015-0142-2.

Netam RS, Tiwari RKS, Bahadur AN, Kumar P, Yadav SC. Efficacy of bio-control agents for the management of (*Pyricularia grisea*) blast disease of finger millet under field condition of Bastar, Chhattisgarh. J Pure Appl Microbiol. 2016;10(3):2421–5.

Ou SH. Blast in: Rice diseases. 2nd ed. Kew, UK: The Commonwealth Mycological Institute; 1985. p. 109–201.

Panwar P, Jha AK, Pandey PK, Gupta AK, Kumar A. Functional markers based molecular characterization and cloning of resistance gene analogs encoding NBS-LRR disease resistance proteins in finger millet (*Eleusine coracana*). Mol Biol Rep. 2011;38:3427–36.

Rossman AY, Howard RJ, Valent B. *Pyricularia grisea* the correct name for the rice blast disease fungus. Mycologia. 1990;2:509e12.

Saccardo PA. Conspectus generum fungorum italiae in feriorum. Michelia. 1880;2:1–135.

Sreenivasaprasad S. Finger millet blast in East Africa: pathogen diversity and disease management strategies. DFID Crop Protection Programme, Final Technical Report, Project R8030. UK: Horticultural Resources International; 2004. 86 p.

Todman AK, Pawar DR, Joshi MH. Host reactions to finger millet blast (*Pyricularia grisea* Sacc.). Mysore J Agric Sci. 1994;28:45e6.

Viji G, Gnanamanickam SS, Levy M. DNA polymorphisms of isolates of *Magnaporthe grisea* from India that are pathogenic to finger millet and rice. Mycol Res. 2000;104(2):161e7.

Vishwanath S, Sanne GS, Seetharam A, Gowda SBT. Reaction to blast disease of released and pre-released varieties of finger millet from different states. Millet News Let. 1986;5:31.

Chapter 5
Chemicals for the Management of Paddy Blast Disease

C. Amoghavarsha, D. Pramesh, E. Chidanandappa, H. Sharanabasav, A. Raghunandana, M. K. Prasanna Kumar, Rajashekara Hosahatti, and A. Saddamhusen

Contents

C. Amoghavarsha
Department of Plant Pathology, University of Agricultural and Horticultural Sciences, Shivamogga, Karnataka, India

D. Pramesh (✉) · E. Chidanandappa · H. Sharanabasav · A. Raghunandana
Rice Pathology Laboratory, All India Coordinated Rice Improvement Programme, University of Agricultural Sciences, Raichur, Karnataka, India
e-mail: pramesh84@uasraichur.edu.in

M. K. Prasanna Kumar · A. Saddamhusen
Department of Plant Pathology, University of Agricultural Sciences, GKVK, Bengaluru, Karnataka, India

R. Hosahatti
ICAR-Vivekananda Parvatiya Krishi Anusandhan Sansthan, Almora, Uttarakhand, India

S. C. Nayaka et al. (eds.), *Blast Disease of Cereal Crops*, Fungal Biology, https://doi.org/10.1007/978-3-030-60585-8_5

5.1 Introduction

Chemicals are vital elements of effective disease management programs. Over the past 200 years, fungicides are being used to protect the plants from fungal diseases. Even though from the beginning there is an increase in the cultivation of crops and the treatment of these crops for the disease management within the limited range of available chemicals, the number of applications and bio-efficacy of chemicals have significantly increased, especially after World War II. In the late 1960s and 1970s, a large number of efficient fungicides with systemic function and new structures, which were not available in the previous products, were introduced. These chemicals consist of benzimidazoles, carboxanilides, dicarboximides, sterol demethylation inhibitors (DMIs), morpholines, 2-amino-pyrimidines, phosphorothiolates and phenylamine. However, in the late 1980s new fungicides were released which were corresponding to the existing fungicides belonging to the DMIs exhibiting enhanced bio-efficacy properties and more environmentally friendly. Over the last decade, a number of new chemical compounds such as aniline pyrimidines, benzamides, carboxylic acid amides, phenyl-pyrroles and quinone outside inhibitors (QoIs, strobilurin analogues) were commercially introduced for plant disease management (Brent and Hollomon 2007).

Rice is an important cereal crop of the world which feeds two-thirds of the world population. Like other cultivated crops, rice production is also limited by the number of biotic problems which include blast, a destructive fungal disease. Rice blast caused by an ascomycetous fungi *Magnaporthe oryzae* (anamorph, *Pyricularia oryzae*) is the most destructive disease on rice crop in most of the rice-cultivating ecosystems of the world. The most common ways followed for the rice blast management under field conditions include manipulation of the sowing time or date of transplantation, cultivation of resistant varieties or hybrids, application of chemical fungicides, controlled fertilisers especially nitrogenous, and irrigation schedules (Mbodi et al. 1987; Moletti et al. 1988; Naidu and Reddy 1989; Georgopoulos and Ziogas 1992). Among various strategies employed for the management of blast disease, chemical-based control has been widely adopted in many countries (Mariappan et al. 1995). Seed treatment and foliar application with the systemic fungicides were found to be very effective in reducing the blast disease (Manandhar et al. 1985; Sah and Karki 1988; Manandhar 1984; Chaudhary and Sah 1997; Chaudhary 1999).

In the early days, introduction of organo-mercuric compounds for seed treatment was later extended for the field application for controlling rice blast. This practice greatly contributed to the protection of rice from blast disease which helped in the improved production of rice. But the report of mercury poisoning causing the 'Minamata disease' in Japan in the year 1956 changed the chemical use pattern for the management of paddy blast. Even though the later findings proved that the 'Minamata disease' was caused by industrial wastes, the use of organic mercuric chemicals for plant disease control was prohibited, and over time, non-mercuric fungicides with a superior mode of action were developed against the blast disease of rice (Yamaguchi 2004).

Several anti-blast chemicals have been developed over time which can be grouped into the following types based on their origin and mode of action:

Fungicidal chemicals (which directly affect the fungal physiology):

- Fungicides of microbial origin
- Anti-blast chemicals inhibiting sulfhydryl enzyme
- Anti-blast chemicals inhibiting cell division
- Sterol biosynthesis inhibitors (SBIs)
- Anti-blast chemicals inhibiting the membrane phospholipid biosynthesis
- Fungicides targeting fungal membrane permeability
- Fungicides suppressing fungal respiration
- Combination fungicides with multiple modes of action

Non-fungicidal chemicals (which indirectly interfere with disease development):

- Melanin biosynthesis inhibitors (these chemicals inhibit only infection process and are not directly lethal to the fungal pathogen)
- Inducers of the systemic acquired resistance (SAR)

5.2 Fungicidal Chemicals Directly Affecting the Pathogen Physiology

This group includes the molecules which are lethal to fungi by inhibiting one or more physiological processes in the fungal cell. These molecules directly affect the pathogen and thereby control the disease caused by them.

5.2.1 Fungicides of Microbial Origin

(a) Blasticidin S

Blasticidin S was the prime agricultural antibiotic which was developed successfully in Japan from the actinomycetes (*Streptomyces griseochromogenes*) by isolating from the filtrates of this actinomycete culture in the year 1958 by Takeuchi and associates. Later its effect in curing the rice blast disease was found, and the evaluation of this chemical was conducted during 1959–1960 through field trials.

This antibiotic gives excellent control of rice blast disease upon spraying at low concentration (10–20 ppm). It inhibits the protein synthesis in rice blast fungi (Misato et al. 1959) by interfering the ribosomes during the peptidyl transfer (Yukioka et al. 1975). At a higher dose, this antibiotic causes mild-to-moderate phytotoxic effect on rice and other crop plants. Benzyl-amino-benzene-sulfonate, a derivative of blasticidin S, was established with low phytotoxicity to plants without compromising the disease control efficacy, and this has been produced commercially for the practical use.

The inhibitory action of blasticidin S is mainly on mycelial growth of the fungus and was found to be 10–100 times more powerful than that of organomercuric fungicides. Therefore, it gives excellent disease control, especially against neck blast. The residues of blasticidin S in the environment are less persistent and easily broken down by sunlight and also by microorganisms in the soil.

Blasticidin S exhibited the prominent blast control and was found to be effective as a curative fungicide. But it is an inferior protectant compared to modern fungicides with extreme toxicity to cultivated plants (Ou 1985).

(b) Kasugamycin

It is also an actinomycetous antibiotic that belongs to the aminoglycoside group of antibiotics isolated from *S. kasugaensis* in 1965 from the soil in Japan by Umezawa and associates. Though there is a nonsignificant in vitro antifungal activity of kasugamycin, it has exhibited the curative action for rice blast under field condition. This is mainly because the hydrogen ion concentration in the tissue of rice plants is around pH 4.5–5.0 which favours the inhibition of the fungal growth by kasugamycin that requires low pH for its activity. Generally, the antibiotics of the group aminoglycosides act by inhibiting protein synthesis by misreading of codons, while on the contrary kasugamycin was found to specifically interfere with the initiation complex formation by preventing the aminoacyl-tRNA binding to RNA-30S ribosomal subunit complex.

Kasugamycin is effective on rice blast and has been widely used in agriculture since 1965 (Ishiyama 1965; Singh et al. 2014). This was the major fungicide used for the control of paddy blast with 4–5 sprays per cultivation season during the early 1970s, and this antibiotic accounted for 90% of the chemicals that were applied for the paddy blast control (Miura et al. 1975). Use of kasugamycin on rice also has added advantage of low or no phytotoxicity on rice at higher doses. After the repeated application of this antibiotic intensively for several years, its efficacy was reduced due to development of resistant field strains. Afterwards, kasugamycin was used in combination with other chemical fungicides with the varied mode of action. Fortunately, it has been observed that the proportion of resistant strains declined rapidly after discontinuing the sole use of kasugamycin for the management of blast.

5.2.2 Anti-blast Chemicals Inhibiting Sulfhydryl Component of Cellular Enzyme

The group includes the complex fungicides, ethylene bis-dithiocarbamates (EBDCs)/dithiocarbamates, that were used for managing various diseases from late 1940s, exclusively in the complex form with zinc (zineb) or manganese (maneb) or a combination of manganese and zinc (mancozeb) (Morton and Staub 2008). Among them, zineb and mancozeb have been used for managing blast disease of rice.

(a) **Zineb:** It is a polymeric complex of zinc with a dithiocarbamate. It is a protective fungicide that controls a broad range of diseases. The chemical is fungitoxic only when it is exposed to air, where on exposure it is converted into an isothiocyanate, and thereby inactivates enzymes of the fungi bearing sulfhydryl (SH) groups. This also functions by disturbing the enzyme activity within fungi whenever the metal exchange between zineb and enzymes of fungi occurs. Effectiveness of zineb in controlling blast disease under moderate-to-severe disease severity is well reported (Singh et al. 2014).

(b) **Mancozeb:** It is a protective contact fungicide with multiple sites of action. It is developed by combining two dithiocarbamates, viz. zineb and maneb. The mancozeb inactivates enzymes of the fungi bearing sulfhydryl (SH) groups of amino acids within fungal cells, thereby interrupting the respiration, lipid metabolism, and synthesis of adenosine triphosphates (ATPs).

This fungicide is widely used for the management of rice blast throughout the world for its superior efficacy against blast disease. Presently, mancozeb is being used as a component with single-site target curative fungicides in many combination products to enhance bio-efficacy and also to prevent the development of fungicide resistance.

5.2.3 Anti-blast Chemicals Inhibiting Benzimidazole Cell Division

This class of fungicides includes the most extensively used fungicide in agriculture such as carbendazim. It is a broad-spectrum systemic fungicide and effectively controls many fungal pathogens of the phylum Ascomycota and the Basidiomycota. Within plants, the benzimidazoles are systemic or mobile and the fungicide is degraded by the microbes in the soil and water, thereby limiting its soil toxicity. Through the phenomenon of photolysis and hydrolysis, this fungicide is metabolised in the plant system. The fungicide kills the cells of the fungi during the mitosis stage of the cell division by specifically targeting the mitotic spindle microtubule synthesis by binding to the β-tubulin and thereby inhibiting the polymerisation of β-tubulin by interacting with it directly.

(a) **Carbendazim:** It is a systemic and broad-spectrum benzimidazole fungicide. Although its exact mode of action is unknown, carbendazim is found to suppress the microtubule assembly by binding to tubulin at an unspecified site, thereby resulting in the halt of cell cycle at the G2/M phase and initiation of apoptosis. Carbendazim is the member of the class of benzimidazoles (2-aminobenzimidazole) and the primary amino group of benzimidazoles is substituted by a methoxy-carbonyl. For the management of paddy blast, it has been used in different ways, viz. seed treatment, seedling dip and foliar sprays of standing crop.

It is suggested to use other contact fungicides with the carbendazim in an alternate manner or using as tank mix application of carbendazim with contact fungicides to manage the development of fungicidal resistance. The carbendazim in combination with other fungicides gave effective results in managing paddy blast compared to treatment with carbendazim alone (Pramesh et al. 2016a).

5.2.4 Sterol Biosynthesis Inhibitors (SBIs)

Sterols are the important lipid molecules found in plants, animals and fungi naturally. The sterol that is found in the fungal cell membrane is commonly called as ergosterol (ergosta-5, 7, 22-trien-3β-ol). The name ergosterol is derived from disease ergot, from which it was first isolated from the members of genus *Claviceps*. Ergosterol is an important constituent of the cell membrane of yeast and other fungi and serves various similar functions as that of cholesterol in animal cells. Its unique presence in the higher fungi may be because of the exposure of these organisms to harsh climatic situations, viz. extremely fluctuating humidity and moisture in their particular ecological niches like the surfaces or within plants or in soil. Since the ergosterol is found in the fungal cell membrane, its absence in plants and animals makes it a useful target for the development of antifungal chemicals.

Sterol biosynthesis inhibitors (SBIs) are assorted into different FRAC (Fungicide Resistance Action Committee) groups based on their mode of action as G_1, G_2, G_3 and G_4, where each group has compounds exhibiting different target sites in the sterol biosynthesis pathway. The chemicals under the groups G_1, G_2 (amines including spiroxamine) and G_3 (hydroxyanilides) are only used as agrochemicals. The members of the group G_4 are of pharmaceutical importance. They are called as demethylation inhibitors (DMIs) because of the target site of fungicides, which in G_1 being sterol C_{14} demethylase. Most of DMIs belong to the triazole group of fungicides and account for nearly 90% of SBIs.

SBIs are found to exhibit long-lasting broad-spectrum systemic action with protective and curative activity and also has a relatively slow development of field resistance (Kuck et al. 2012).

(a) **Triazole fungicides:** The triazole fungicides are the members of the group demethylation inhibitors (DMIs), which were introduced during the mid of 1970s. Triazoles comprise legion members, among which hexaconazole, tebuconazole and tricyclazole are used for the control of paddy blast. Triazole fungicides act by inhibiting a specific enzyme, C_{14}-demethylase, which is responsible for the production of sterol in the fungal cell membrane. Ergosterol is very much necessary for the structure and function of cell membrane which makes it an essential factor for the functional cell wall development. Application of these compounds leads to the abnormal growth of the fungi, thereby causing the eventual death.

The triazole compounds act on the biochemical pathway of sterol synthesis in a bit different way. Even though it exhibits similar results as with other fungicides of this group leading to abnormal fungal growth and eventual death, the main difference in these fungicides lies in their activity spectra. The triazoles were found ineffective against germination of spores because of the presence of enough sterols in the spores which is sufficient to form germ tubes. Some of the fungal spores were even found to comprise sterol that is adequate to produce infection structures; hence, in some of the fungal diseases, triazoles are ineffective in preventing the host from infection (Mueller 2006).

The application of triazoles can be carried out as early-infection treatments or as a preventive spray. But the applications of the fungicide must be done early during the infection process of the fungi. Few of the fungicides of this group have anti-sporulation characteristics, where the application of such fungicide will inhibit sporulation and thereby slow down the development of the disease. However, the triazole fungicides are ineffective if the fungus starts sporulation on infected plant tissue.

Use of tricyclazole for managing different stages of blast disease of rice has been reported by many research groups (Froyd et al. 1976). Presently it is the most widely used curative fungicide for blast disease management in India.

5.2.5 Anti-blast Chemicals Inhibiting the Membrane Phospholipid Biosynthesis

The organophosphorus compounds were initially introduced as an insecticide, but some of these compounds were found to be effective against the rice blast. The phosphorothiolate compounds having anti-blast activity were introduced in Japan during the year 1963 and they are still one among the major fungicide groups for managing paddy blast. The discovery of phosphorothiolate (PTL) compounds with antifungal activity has led to the development of fungicides, viz. iprobenfos (IBP), edifenphos (EDDP) and isoprothiolane, which were used at a greater extent for the management of rice blast.

(a) **IBP:** It is an organophosphorus compound with high water solubility and systemic action within the plant, which favours the use of this fungicide as soil and water surface application in the paddy fields.

(b) **EDDP:** It is a non-systemic, organophosphorus compound with impressive antifungal activity against the rice blast pathogen. It acts as an efficient blast control fungicide when used as a foliar spray because of its less solubility in water and less stability in plants (Uesugi 2001).

    IBP and EDDP specifically act by inhibiting the conversion of phosphatidylethanolamine into phosphatidylcholine (Kodama et al. 1979). Phospholipid is an important constituent of a fungal cell membrane; inhibition of the synthesis of phosphatidylcholine causes impairment of membrane permeability and affects the associated enzyme activities. Among various systemic chemicals analysed, the best blast control was observed with IBP on application to the water surface of paddy (Yamaguchi 1974).

(c) **Isoprothiolane:** It is a systemic organophosphorus compound that primarily acts by preventing the fungus invasion into the plant system (Araki and Miyagi 1976; Taninaka et al. 1976). Even though isoprothiolane (di-isopropyl 1, 3-dithiolan-2-yliden-malonate) is chemically different from the other PTL fungicide compounds, the cross-resistance between isoprothiolane and other PTL compounds intimates resemblances in their mode of action (Katagiri and Uesugi 1977).

The isoprothiolane acts by inhibiting the penetration and elongation of infection hyphae by affecting the formation of infection peg by cellulase secretion. A study conducted by Raji and Louis (2007) revealed that the effectiveness of isoprothiolane against leaf and neck blast of paddy was excellent.

5.2.6 Fungicides Targeting Fungal Membrane Permeability

This group includes chemical ferimzone. It is a systemic fungistatic compound that was developed for the management of paddy blast. The fungistatic ferimzone is not detrimental to the blast pathogen in vitro, but it leads to the leakage of the acidic electrolytes specifically from the *M. oryzae* mycelia. It acts by impairing the membrane permeability and thereby affecting the influx and efflux of salt ions or specific molecules. Because of its fungistatic nature, ferimzone is used as a mixture with other fungicides as a synergistic component (Okuno et al. 1989). Ferimzone-containing two products, Blacin and Nonblas, were registered in 1991 in Japan and released into the market. Blacin contains fthalide in addition to ferimzone and Nonblas contains tricyclazole with ferimzone to escape resistant strain development and to provide more effective blast control (Matsuura et al. 1994).

Leaf blast field trials were conducted in Japan by using the ferimzone and it provided good control of blast in each application. It was observed that the ferimzone has high curative activity against the disease (Matsuura et al. 1994).

5.2.7 Fungicides Suppress Fungal Respiration

This class includes the important fungicides which target mitochondrial respiration of fungi. Sauter et al. (1995) proposed the usage of the term "strobilurins" for the members either occurring naturally or produced synthetically for this class of fungicides due to the natural molecule strobilurin A, from which the fungicides are derived with structural variations. Strobilurins A and B were the first extracted natural compounds from the fungus *Strobilurus tenacellus*. These compounds were originally known as β-methoxyacrylates, or β-MOAs, because of its toxophore structure, which is an active part of the molecule that is responsible for its toxic effect. Eventually, different molecules with a similar mode of action were synthesised with a toxophore structure analogous to strobilurin A, but not that of β-MOA.

A variety of strobilurin compounds are currently used in agriculture for managing many fungal diseases. The bc1 complex of the mitochondrial respiration is involved in transferring electrons from ubiquinol to the cytochrome c oxidase. These chemicals block the electron transfer chain at the ubiquinol-oxidising site (Qo) in the inner mitochondrial membrane and thereby inhibiting the ATP synthesis and cellular respiration (Shen et al. 2014).

The naturally occurring strobilurins were less active and photo-insensitive; hence the strobilurins were optimised for the fungicidal activity and photostability, and developed commercially. They were first released in 1996 and the current market has more than ten strobilurin fungicides with 23–25% fungicide sales globally. Several strobilurins, viz. azoxystrobin, kresoxim-methyl, metominostrobin, picoxystrobin, pyraclostrobin and trifloxystrobin, were developed commercially and used for the paddy blast management.

(a) **Azoxystrobin:** It is a broad-spectrum methoxyacrylate class of fungicide with a systemic activity which is a derivative of the naturally occurring strobilurin. It is found very effective in managing paddy blast on its application during heading stage of the crop. However, the common trend of a single application of this fungicide in the late crop season to manage both sheath blight and blast has led to less efficient control of rice blast (Groth 2006). Nowadays, this molecule is also being used as a component of a combination product with the triazole group of fungicides.

(b) **Kresoxim-methyl:** It is a systemic and broad-spectrum fungicide exhibiting both protective and curative activity. It has good residual activity apart from the broad-spectrum disease control and hence imparts extended duration of disease control. Kresoxim-methyl is effective even at very low concentrations and inhibits the spore germination of the fungus on the host tissue, thereby preventing the infection and spread of the disease. It protects from a broad range of fungal diseases and has a primary role in plant disease management.

 It can cure the already advanced infections and halts the sporulation and symptom expression, thereby impeding further advancement of the disease. These characteristics make kresoxim-methyl an excellent compound in integrated disease management (IDM) programs. For the management of paddy blast, kresoxim-methyl showed higher protective and curative performance against *M. oryzae* isolates exhibiting excellent efficacy (Chen et al. 2015).

(c) **Metominostrobin:** It is a broad-spectrum methoxyacrylate fungicide with systemic activity. It acts by inhibiting the respiration of fungi by blocking the electron transport in the cytochrome-bc1 segment of the respiratory chain in the inner membrane of mitochondria. However, the cells of *M. oryzae* mycelia induce cyanide-resistant respiration to resume the respiratory cycle on the blockage of the cytochrome-mediated pathway by metominostrobin. The superoxide anion is thought to be responsible for the induction of the cyanide-

resistant respiration mechanism involving the metominostrobin-dependent induction. Flavonoids are water-soluble polyphenols found in plants and they are capable of scavenging the superoxide anions that were produced due to the obstruction of the electron flux via the section of cytochrome-bc 1 complex and thereby inhibiting the cyanide-resistant respiration induced by the metominostrobin activity. In this way, the metominostrobin successfully manages the rice blast pathogen with the help of host plant factors (Mizutani et al. 1996).
Metominostrobin exhibits best disease-controlling activity immediately after application against rice blast. Furthermore, it shows long-lasting activity when applied to paddy water. It can control not only leaf blast but also ear blast even at 60 days after submergence (Mashiko et al. 2001; Gaikwad and Balgude 2016).

(d) **Picoxystrobin:** It is a broad-spectrum strobilurin compound with both preventive and curative nature and acts by inhibiting fungal respiration. This compound exhibits better curative property in comparison to the azoxystrobin in many crop diseases.
(e) **Pyraclostrobin:** It is a systemic and broad-spectrum fungicide exhibiting excellent fungicidal properties with translaminar and loco-systemic activity. It is curative, protective and eradicative which makes it better in comparison to other fungicides. It is absorbed rapidly by plants and retained largely in the leaf cuticle by the waxes. Because of its translaminar property, it gives the best control of the disease on both the leaf surfaces. It has a very confined vapour phase and acropetal and basipetal leaf movement activity. The preventive spray with the concentration as low as 0.1 ppm successfully suppresses the spore germination of a majority of pathogens.
Pyraclostrobin was assessed for its efficacy and found to be most effective by recording least percent disease index (PDI) for leaf blast disease (Pramesh et al. 2016b).
(f) **Trifloxystrobin:** This has been successfully used as a component of combination product along with other groups of fungicides such as triazoles. From the 1990s, this fungicide was extensively used for the effective management of fungal diseases of various crops (Liu et al. 2013). It is grouped as highly toxic to non-target aquatic species but as non-toxic to bees, agriculturally important insects, earthworms and mammals (Junges et al. 2012; Shen et al. 2014).

5.2.8 Anti-blast Chemicals with Combined Mode of Action

Many fungicide combinations with different modes of action have been developed for the management of paddy blast. The important combination products that were found effective against paddy blast include difenoconazole 11.4% + azoxystrobin 18.2% SC, mancozeb 63% + carbendazim 12% WP, copper oxychloride 45% + kasugamycin 5% WP, picoxystrobin 6.78% + tricyclazole 20.33% SC, tricyclazole 18% + mancozeb 62% WP and tebuconazole 50% + trifloxystrobin 25% WG.

(a) **Difenoconazole 11.4% + azoxystrobin 18.2% SC:** It is a combination fungicide containing difenoconazole and azoxystrobin. It exhibits a dual-systemic nature with broad-spectrum fungicidal activity with protective and curative action. It acts by inhibiting the spore germination at an early stage of pathogen development. Thus, it protects the crop against invasion by fungal pathogens. Once it is absorbed by plants systemically, it inhibits the fungal penetration and haustoria formation by obstructing the sterol biosynthesis in the fungal cell membrane and thereby preventing the disease development. Singh et al. (2019) reported the use of azoxystrobin 18.2% + difenoconazole 11.4% SC for the paddy blast with the best control of the disease.

(b) **Copper oxychloride 45% + kasugamycin 5% WP**: It is a new combination product containing copper oxychloride (COC) and kasugamycin, which has the power of fungicide and bactericide together to prevent the bacterial-fungal complex in multiple crops. Its dual mode of action makes it an effective and powerful tool to protect crops from fungal and bacterial diseases. It is a contact and systemic fungicide which interferes with the enzyme system of spores and mycelium and inhibits protein biosynthesis.

A spray with a combination of COC and antibiotic (copper oxychloride 45% + kasugamycin 5% WP) was found effective against leaf blast among different combinations (Kumar and Veerabhadraswamy 2014), and suitable for the management of paddy blast under field conditions.

(c) **Carbendazim 12% + mancozeb 63% WP:** It is a combination fungicide having a mixture of carbendazim and mancozeb. One of the components, mancozeb, remains on the surface of the plant and prevents the development of the disease by acting as the contact fungicide. Whenever mancozeb is exposed to air, it becomes fungitoxic due to the conversion into an active isothiocyanate that acts by inactivating the fungal SH (sulfhydryl) groups of enzymes and thereby disturbing the functioning of enzymes in the fungal cell. Another component of the mixture, carbendazim, is absorbed by the plant due to its systemic activity and protects the plant from invading pathogen and also acts as a curative agent. It inhibits the fungal germ tube development and appressoria formation, thereby preventing the growth of mycelia.

The use of the fungicide combination involving carbendazim 12% and mancozeb 63% WP against the paddy blast has resulted in less disease incidence (Pramesh et al. 2016a).

(d) **Picoxystrobin 6.78% + tricyclazole 20.33% SC:** It is a combination of fungicides containing picoxystrobin 6.78% + tricyclazole 20.33% SC with dual mode of action, both preventive and curative for effective management of blast disease of paddy. It is an effective systemic fungicide that gets rapidly absorbed and translocated all over the plants. After spraying, the chemical combination is absorbed rapidly by the plant and moves towards leaf tips. The fungicide acts by effectively blocking the penetration of fungus on germination, and thereby preventing the establishment of the pathogen at the infection court. Tricyclazole-treated spores are unable to penetrate as it cannot synthesise melanin and by doing so cannot generate enough turgor pressure to rupture host cuticle.

Tricyclazole also inhibits spore formation and release from sporophores and whenever the spores are formed, they are less virulent.

(e) **Tebuconazole 50% + trifloxystrobin 25% WG**: It is a new combination fungicide containing tebuconazole 50% + trifloxystrobin 25% WG. This combination is a broad-spectrum and systemic fungicide that is having both protective and curative properties and along with the disease control, it also improves the quality and crop yield. In rice, this combination also protects against the dirty panicle disease and the sheath rot in later stages of the crop. Tebuconazole is basically a demethylase inhibitor (DMI) and interferes during the process of the formation of fungal cell wall structures, thereby stopping the fungal growth and reproduction. Trifloxystrobin is a strobilurin fungicide and acts by impairing the fungal respiration by interfering in the mitochondrial electron transport chain.

The evaluation of different fungicide combination was carried out and it was found that fungicide trifloxystrobin 25% + tebuconazole 50% performed better with least disease incidence over other chemicals (Pramesh et al. 2016b).

(f) **Tricyclazole 18% + mancozeb 62% WP**: It is a combination fungicide with both systemic and contact activity and contains the tricyclazole and mancozeb. The systemic part of tricyclazole on spraying is absorbed rapidly by the plant and moves towards leaf tips. The fungicide acts by effectively blocking the penetration of the fungus on germination and thereby preventing the establishment of the pathogen at the infection court. In the rice blast disease, melanin is required for the hardening of appressorium, and whenever the pigment formation is affected, the formed appressoria fail to successfully penetrate the host surface. Another compound in the combination, mancozeb, is a broad-spectrum contact and protective fungicide. Whenever the mancozeb is exposed to air it becomes fungitoxic due to the conversion into an active isothiocyanate that acts by inactivating the fungal SH (sulfhydryl) groups of enzymes, and thereby disturbing the functioning of enzymes in the fungal cell.

Chethana (2018) evaluated the efficacy of different fungicides and their combination for the rice neck blast and found that the fungicide combination containing tricyclazole and mancozeb performed better by effectively reducing the disease incidence.

5.3 Non-fungicidal Chemicals to Control Rice Blast Disease

Majority of the chemicals that were used for the management of paddy blast are non-fungicidal and prevent the disease development by acting on fungal metabolism, viz. melanin biosynthesis, or by inducing the defence mechanism in plants, such as systemic acquired resistance (SAR). They are non-lethal to fungal cell but interfere with the secondary metabolite synthesis during plant infection and disease development, and thus help to control the disease.

5.3.1 Melanin Biosynthesis Inhibitors (MBIs)

A class of chemical compounds that are known to block specifically the melanin biosynthesis pathway in the pathogen secondary metabolism are called melanin biosynthesis inhibitors (MBIs). These chemicals have been used for a long time for the management of paddy blast (Hamada et al. 2014).

Melanin is high-molecular-weight, black- or brown-coloured pigment which is negatively charged, hydrophobic and formed from the indolic and phenolic compounds by oxidative polymerisation (Wheeler and Bell 1998). Melanin protects the microorganism from different harsh and toxic environments, and hence acts as the fungal armour. It protects the fungi from the ultraviolet rays (UV), extreme environmental conditions, enzymatic lysis, desiccation, antimicrobial drugs, phagocytosis, oxidants and heavy metal ions (Pal et al. 2013). The fungus is known to synthesise melanin either via L 3–4 dihydroxyphenylalanine (L-DOPA) pathway which is found in Basidiomycota or 1,8-dihydroxynaphthalene (DHN) pathway that is usually seen in Ascomycota (Bell and Wheeler 1986).

Melanin is required for the appressorium to penetrate host surface. Accumulation of melanin in the walls of appressorium improves the rigidity and cell capacity to withstand high turgor pressure that is developed during the penetration of host surface (Woloshuk et al. 1980, 1983). The accumulation of DHN melanin is at the inner layer of the appressorial cell wall which on accumulation changes the porosity of the cell wall, thereby preventing the efflux of larger molecules from the cell with increased osmotic gradient due to the enhanced intracellular glycerol concentration. The increased osmotic gradient will result in an accelerated influx of water molecules into the cell leading to the generation of high turgor pressure (>8.0 MPa), which enables the physical penetration of hard barrier like plant surface (Howard et al. 1991).

Magnaporthe produces a dark brown melanin pigment via penta-ketide pathway from 1,8-dihydroxynaphthalene (1,8-DHN) in order to overcome the host self-defence mechanism; the pathogen has developed an infection process that successfully causes disease in plants. Whenever the germinating fungal hyphae identify the host surface characteristics, viz. lipophilicity and hardness, the hyphae differentiate into appressoria through which it infects the harder host surface. The appressoria formation involves the expression of many genes and signal transduction within the pathogen.

For successful penetration of the host, the melanised appressoria are very much essential. Conidia of the fungi germinate and form germ tube, where the tips of the germ tube differentiate into appressoria. The formation of melanin layer in between the cell wall and the plasma membrane leads to maturation of the appressoria that is capable to penetrate the host cuticle by producing a high osmotic pressure within the cell. Hence, the inhibitors of the melanin biosynthesis can effectively control rice blast disease development (Yamaguchi and Kubo 1992).

Melanin synthesis inhibitors, viz. carpropamid, fthalide, isoprothiolane, pyroquilon and tricyclazole, exhibit excellent control against blast disease even though

Table 5.1 Melanin biosynthesis inhibitor groups

Inhibitors group	Chemical name	Common name
Reductase inhibitors (MBI-R)	Polychlorinated aromatic compounds	Fthalide (FTL)
	Fused heterocyclic compounds	Tricyclazole (TCZ), Pyroquilon (PQR)
Dehydratase inhibitors (MBI-D)	Carboxamide derivatives	Carpropamid (CAR), Diclocymet (DCM), Fenoxanil
Polyketide synthase inhibitors (MBI-P)	Trifluoro-ethyl-carbamate	Tolprocarb

they are non-toxic to the growth of *M. oryzae* mycelia. The main reactions in DHN melanin biosynthesis pathway involve reduction and dehydration by enzymes. Therefore, the enzymes, viz. reductase and dehydratase, act as targets for producing inhibitors of melanin biosynthesis (Table 5.1).

Melanin biosynthesis inhibitors (MBIs) targeting scytalone dehydratase (MBI-D) (called as dehydratase inhibitors) were invented in the year 1998 as specific and efficient fungicides to manage rice blast. These fungicides were released as granule fungicides for the treatment of a nursery box. The usage of MBI-D fungicide compounds unfolded rapidly due to the numerous advantages of box treatment nursery, viz. reduced quantity and frequency of application, effectiveness over a long time and reduced working hours (Suzuki et al. 2010).

However, during the year 2001, it was reported that MBI-D failed to manage rice blast and the resistant isolates to MBI-D were confirmed and their use was discontinued as a countermeasure to overcome developing resistance. After 2003, the use of other types of MBI targeting polyhydroxy-naphthalene reductase (MBI-R) fungicide (called as reductase inhibitors) was increased and it restored the level of use of MBIs which subsisted before the release of MBI-D fungicides in 1998 (Suzuki et al. 2010).

5.3.1.1 Reductase Inhibitors

These MBI act by inhibiting the two reduction steps between 1,3,6,8-trihydroxy-naphthalene (1,3,6,8-THN) and scytalone and between 1,3,8-THN and vermilion (Chida and Sisler 1987b). Fthalide (FTL), tricyclazole (TCZ) and pyroquilon (PRQ) are the members of this group. These were introduced earlier in granular form and applied to the irrigated paddy by submerged application or by nursery box treatment.

(a) **Fthalide**: It is an excellent protectant fungicide with extended residual activity. Its primary metabolites are found to have negligible phytotoxicity and mammalian non-toxicity on detailed examination for its behaviour and metabolic fate (Tokuda et al. 1976). Fthalide application inhibits the melanin biosynthesis in the fungal appressoria, thereby interfering the host penetration by the pathogen (Chida and Sisler 1987a).

(b) **Pyroquilon**: It has a similar mode of action as that of other melanin biosynthesis inhibitors but with an accumulation of scytalone and 2-hydroxyjuglone (2-HJ) predominantly (Yamaguchi et al. 1982). The application of pyroquilon and tricyclazole induces the accumulation of flaviolin at high concentration, thereby indicating inhibition at another step, which is catalysed by the similar enzyme. This fungicide reduces sporulation of blast fungus and thereby controls the secondary infections under field conditions. Results of the field studies of tricyclazole and pyroquilon have shown the long-term disease control property when applied as either a soil drench, foliar spray or submerged application.
A study revealed that seed treatment with pyroquilon in susceptible cultivars keeps leaf blast severity very low. Seed treatment at moderate rates with pyroquilon has been found to reduce the severity of leaf blast under low disease pressure. When a susceptible cultivar was used under high disease pressure, the maximum disease control was observed with the application of this fungicide at higher rates. Seed treatment of more resistant cultivars with pyroquilon had no or negligible effect. Thus, the study has shown that the use of pyroquilon for seed treatment gives satisfactory control during the vegetative phase of rice against leaf blast in susceptible cultivars (Prabhu and Filippi 1993).

(c) **Tricyclazole**: It is a systemic fungicide with protective action and controls rice blast by inhibiting the melanin biosynthesis pathway in the appressorial wall. The fungicide acts by effectively blocking the penetration of the fungus on germination and thereby preventing the establishment of the pathogen at the infection court. Tricyclazole affects neither germination of the spore and appressorial formation nor the mycelial growth but particularly affects fungal penetration by suppressing the melanin biosynthesis. The fungicide inhibits production of the late intermediate in melanin pathway, a vermelone, even at the extremely low concentration (Tokousbalides and Sisler 1978; Woloshuk et al. 1980).

5.3.1.2 Dehydratase Inhibitors

These MBIs act by inhibiting the dehydration at two steps, between scytalone and 1,3,8-trihydroxy-naphthalene (1,3,8-THN) and between vermelone and 1,8-dihydroxy-naphthalene (1,8-DHN). The use of reductase inhibitors for a long time led to the delayed discovery of these compounds for the control of rice blast. A few dehydratase inhibitors were introduced in the late 1990s, viz. carpropamid, diclocymet and fenoxanil.

(a) **Carpropamid:** It is a protectant fungicide with systemic activity and was found to be effective against leaf and panicle blast. The fungicide acts by inhibiting the melanin biosynthesis by particularly targeting the dehydratase enzymes and thereby preventing the dehydratase reaction between scytalone and trihydroxy-naphthalene and also vermelone and dihydroxy-naphthalene. In comparison to the older melanin biosynthesis inhibitors, viz. pyroquilon, phthalide and tricyclazole (reductase inhibitors), the carpropamid fungicide has a different site of action which acts by inhibiting the 1,3,8-trihydroxy-naphthalene reductase (Yamaguchi and Kubo 1992).

Kurahashi et al. (1996, 1997) examined the basic activity and biological properties of carpropamid and it was found to be an excellent penetrant for controlling blast. It showed weak activity against oomycetes fungi and it does not inhibit blast fungus spore germination and appressorium formation but even at very low concentration, it strongly inhibits melanisation of appressorium.

5.3.1.3 Polyketide Synthase Inhibitors

(a) **Tolprocarb:** It is a novel systemic chemical used for the management of rice blast. The site of action of this fungicide is polyketide synthase (PKS) that regulates the synthesis of polyketide and cyclisation of penta-ketide in melanin biosynthesis pathway (Hamada et al. 2014). In addition to inhibiting melanin biosynthesis, tolprocarb also induces systemic acquired resistance in rice.

The application of tolprocarb accelerated genes involved in signalling pathway mediated by the salicylic acid, viz. chitinase-1, β-1,3-glucanase and PBZ1, without accelerating genes related to the signalling pathway mediated by the jasmonic acid. Hence, it was found that tolprocarb also induces the systemic acquired resistance (SAR) in rice plants by accelerating the salicylic acid-mediated signalling pathway (Hagiwara et al. 2019).

The tolprocarb fungicide can be used effectively for the control of rice blast which in turn provides resistance against other pathogens by activating the SRA. Because of the dual mode of action of tolprocarb, the risk associated with the development of resistant isolates is very low.

5.3.2 *Inducers of Systemic Acquired Resistance (SAR)*

Plant activators or priming effectors are the groups of chemical compounds which are non-fungicidal in action and effectively control rice blast by inducing systemic acquired resistance (SAR) in plants. Induced resistance enables the plants to improve their level of basal resistance against the invading pathogen. Various biotic and abiotic stimuli assist in the activation of induced resistance (Pieterse et al. 2012). Two types of such induced resistance can be seen in plants, viz. systemic acquired resistance (SAR) and induced systemic resistance (ISR). Both these resistance types vary based on the signalling pathway involved in the activation and nature of the elicitor molecules involved (Knoester et al. 1999). The activation of systemic acquired resistance results in signal transduction via salicylic acid (SA)-mediated pathway, thereby producing the PR (pathogenesis-related) proteins (Knoester et al. 1999). Contrastingly, the non-pathogenic microorganisms activate the induced systemic resistance (ISR) that acts through the pathway mediated by jasmonic acid (JA) and ethylene (ET) where there is no accumulation of PR proteins (Knoester et al. 1999).

Many of such chemicals acting as plant activators against rice blast were developed and used for the practical control of the disease. The most commonly used plant activators or defence inducers include acibenzolar *S*-methyl, probenazole, isotianil and tiadinil. Even though plant activators have no fungicidal activity, they protect plants by activating the resistance mechanism mediated by SA pathway with the accumulation of pathogenesis-related protein (PR protein) (Arie and Nakashita 2007).

(a) **Probenazole (PBZ):** It is an effective systemic compound that effectively controls rice blast on application to the root system (Watanabe et al. 1977). The augmentation of the enzyme activity related to the plant resistance was found at the site of pathogen invasion on the application of PBZ to rice plants. The host-mediated defence activity of this chemical compound improves its efficacy for the extended period.
Minami and Ando (1994) studied the activity of this chemical for the management of paddy blast and found that probenazole successfully induced resistance in the host plant against the invading *M. oryzae* pathogen. Formation of hypersensitive response (HR) lesions was observed on pretreatment of probenazole to plants due to high activity of SA and accumulation of PR proteins.
(b) **Acibenzolar-*S*-methyl (BTH):** It is a plant defence activator and acts by inducing SAR in plants (Yamaguchi 1998). Even though the structures of BTH and PBZ look chemically similar they are found to have a different site of action. BTH acts by inducing the SAR downstream of the salicylic acid, whereas the active metabolite (1,2-benzisothiazole-l, 1-dioxide; BIT) of PBZ acts at a prior step to the salicylic acid (Yoshioka et al. 2001; Nakashita et al. 2002).
It works as a preventive spray and needs to be applied prior to commencement of the disease. The best results were obtained with the spray covering the entire plant surface uniformly despite its systemic nature. To ensure the uniform spray coverage, it can be best applied under sufficient water situations by the aerial spray or through ground application. The BTH gives maximum control of rice blast after 4 days of the application and it was found to mimic SAR in plants.
(c) **Tiadinil**: It is another priming effector and a novel systemic chemical compound that effectively controls rice blast (Umetani et al. 2003). Within the leaf sheath in inner epidermal tissue of rice, tiadinil acts by inhibiting the growth of pathogen hyphae at the first invaded cell by enhancing the deposition of callose in the invaded cells and thereby hindering the development of hyphae. All these functions and the invaded cell's cytoplasmic reaction were found to be similar as observed in resistance reactions during the incompatible host-pathogen interaction.
In the tiadinil-treated plants, the improved expression of genes related to the host resistance, viz. PAL-ZB8, RPR-1 and PBZ1, has been reported, thereby suggesting that tiadinil acts as a plant activator and gives excellent control of rice blast by activating the genes responsible for host resistance against the

invading pathogen. It can be effectively used for the management of blast by water application and nursery box treatment (Tsubata et al. 2006).

(d) **Isotianil:** It is a resistance-inducing chemical belonging to the isothiazole class which stimulates the natural defence mechanisms of rice plants. This provides resistance with low application rates. Isotianil can effectively control the fungicide-resistant strains of blast fungus as it is an inducer of resistance in host plants to invading pathogens. Isotianil exhibits some excellent characteristics and this chemical can be applied to plants by employing different methods, and even with lower dosage, the effect of this chemical is long-lasting in comparison to the existing other plant activators.

In a preventive spray test on rice plants against blast, the inoculation of the pathogen 5 days after the chemical spray exhibited better control in comparison to the inoculation of the pathogen on the next day of chemical spray. This indicated the requirement of spray of isotianil at least a few days before the landing of the pathogen spores on host surface (Sakuma et al. 2008). The detailed list of fungicides used in the management of blast disease is summarised in Table 5.2.

Table 5.2 List of chemicals used for the management of blast disease

Class	Chemical group	Chemical name	Trade name	Dosage[a] (per litre)
Protein synthesis inhibitors	Enopyranuronic acid antibiotic	Blasticidin-S	Bal-S	10–20 mg
	Hexopyranosyl antibiotic	Kasugamycin 3% SL	Kasu-B	1.5–2.0 mL
Cell division inhibitors	Benzimidazole	Carbendazim 50% WP	Bavistin	1.0 g
Phospholipid biosynthesis inhibitors	Organophosphatic (phosphoro-thiolates)	Edifenphos 50% EC	Hinosan	0.6–0.8 mL
		Iprobenfos 48% EC	Kitazin	1.0 mL
	Dithiolane	Isoprothiolane 40 EC	Fuji-one	0.75–1.0 mL
Sterol biosynthesis inhibitors (SBIs)	Triazole	Tricyclazole 75 WP	Beam	0.6 g
		Tebuconazole 25% EC	Folicur	1.0 mL
		Hexaconazole 5% EC	Contaf	1.0 mL
Respiration inhibitors (QoI)	Oximino-acetates	Kresoxim-methyl 44.3% SC	Ergon	1.0 mL
	Oximino-acetamides	Metominostrobin 20% SC	Oribright	0.8 mL
	Methoxy-acrylates	Picoxystrobin 22.52% SC	Galileo	1.0 mL
	Methoxy-carbamates	Pyraclostrobin 100 g/L CS	Seltima	1–1.5 g
Membrane permeability affectors	Pyrimidinone-hydrazones	Ferimzone	Blasin	2.0 g

(continued)

Table 5.2 (continued)

Class	Chemical group	Chemical name	Trade name	Dosage[a] (per litre)
Sulfhydryl enzyme inhibitors	Mn-bis-dithiocarbamate	Mancozeb 75% WP	Indofil M-45	2.0 g
	Zn-ethylene-bis-dithiocarbamate	Zineb 75% WP	Phytox	1–1.5 g
Combination fungicides	Strobilurin + triazole	Trifloxystrobin 25% + tebuconazole 50% WG	Nativo	0.4 g
	Triazole + strobilurin	Difenoconazole 11.4% + azoxystrobin 18.2% SC	Godiwa-Super	1.0 mL
	Triazole + dithiocarbamate	Hexaconzole 4% + zineb 68% WP	Avtar	1.5 g
	Triazole + validamycin	Hexaconazole 5.00% + validamycin 2.50% SC	Valxtra	1.0 mL
	Benzimidazole + dicarboximide	Carbendazim 25% + iprodione 25% WP	Quintal	1.0 g
	Strobilurin + triazole	Kresoxim-methyl 40% + hexaconazole 8% WG	Ayaan	1.0 g
	Dithiocarbamate + benzimidazole	Mancozeb 63% + carbendazim 12% WP	Companion	1.0 g
		Mancozeb 50% + carbendazim 25% WS	Sprint	1.0 g
	Triazoles	Tricyclazole 45% + hexaconazole 10% WG	Impression	1.0 g
	Triazoles	Tricyclazole 34.2% + propiconazole 10.7%	Filia	1.0 mL
	Dithiocarbamate + triazole	Mancozeb 62% + tricyclazole 18% WP	Merger	1.5 g
	Strobilurin + triazole	Picoxystrobin 6.78% + tricyclazole 20.33% SC	Galileo Sensa	1.0 mL
	Aminoglycoside + copper oxychloride	Kasugamycin 5% + copper oxychloride 45% WP	Conika	1.0 g

(continued)

Table 5.2 (continued)

Class	Chemical group	Chemical name	Trade name	Dosage[a] (per litre)
Melanin biosynthesis inhibitors	Cyclopropane-carboxamide	Carpropamid 27.8% SC	Protega	0.3–0.5 mL
	Trifluoroethyl-carbamate	Tolprocarb	–	–
	Isobenzofuranone	Fthalide 50% WP	Rabicide	1.0 g
	Pyrrolo-quinolone	Pyroquilon	Fongoren	1.0 g
Inducers of SAR	Benzisothiazole	Probenazole	Sportak	200 ppm
	Benzo-thiadiazole (BTH)	Acibenzolar-*S*-methyl	Actigard	200 ppm
	Thiadiazole-carboxamide	Tiadinil	V-Get	250 ppm
		Isotianil	Routine	250 ppm

[a]Dose varies with crop age

References

Araki F, Miyagi Y. Effect of isoprothiolane on the infection process of *Pyricularia oryzae*. Ann Phytopathol Soc Jpn. 1976;42:401–6.

Arie T, Nakashita H. Plant activator. Shokubutsuboeki. Plant Prot. 2007;61(10):531–6.

Bell AA, Wheeler MH. Biosynthesis and functions of fungal melanins. Annu Rev Phytopathol. 1986;24:411–51.

Brent KJ, Hollomon DW. Fungicide resistance in crop pathogens: how can it be managed. Crop Life International; 2007. p. 1–55.

Chaudhary B. Effect of blast disease on rice yield. Nepal Agric Res J. 1999;3:8–13.

Chaudhary B, Sah DN. Effect of promising rice genotypes on leaf blast disease progression. Nepal Agric Res J. 1997;1:27–31.

Chen Y, Yang X, Yuan SK, Li YF, Zhang AF, Yao J, Gao TC. Effect of azoxystrobin and kresoxim-methyl on rice blast and rice grain yield in China. Ann Appl Biol. 2015;166:434–43.

Chethana BS. A new combination fungicide for the management of sheath blight and neck blast diseases of paddy. Int J Pure Appl Biosci. 2018;6(4):651–5.

Chida T, Sisler HD. Restoration of appressorial penetration ability with melanin precursors in *Pyricularia oryzae* treated with antipenetrants and melanin-deficient mutants. J Pestic Sci. 1987a;12:49–55.

Chida T, Sisler HD. Effect of inhibitors of melanin biosynthesis on appressorial penetration and reductive reactions in *Pyricularia oryzae* and *Pyricularia grisea*. Pestic Biochem Physiol. 1987b;29:244–51.

Froyd JD, Paget CJ, Guse LR, Dreikorn BA, Pafford JL. Tricyclazole: a new systemic fungicide for control of *Pyricularia oryzae* on rice. Phytopathology. 1976;66(1):135–1.

Gaikwad AP, Balgude YS. Metominostrobin, a novel strobilurin fungicide for managing rice blast. J Rice Res. 2016;9(1):50–2.

Georgopoulos SG, Ziogas BN. Principles and methods for control of plant diseases. Athens; 1992. 236 p.

Groth DE. Azoxystrobin rate and timing effects on rice head blast incidence and rice grain and milling yields. Plant Dis. 2006;90:1055–8.

Hagiwara H, Ezaki R, Hamada T, Psuda M, Ebihara K. Development of a novel fungicide, tolprocarb. J Pestic Sci. 2019;208–213.

Hamada T, Asanagi M, Satozawa T, Araki N, Banba S, Higashimura N, Akase T, Hirase K. Action mechanism of the novel rice blast fungicide tolprocarb distinct from that of conventional melanin biosynthesis inhibitors. J Pestic Sci. 2014;39(3):152–8.
Howard RJ, Ferrari MA, Roach HD, Money JP. Penetration of hard substrate by a fungus employing enormous turgor pressure. Proc Natl Acad Sci U S A. 1991;88:11281–4.
Ishiyama T. Studies on the preventive effect of kasugamycin on rice blast. J Antibiot. 1965;18:115–9.
Junges CM, Peltzer PM, Lajmanovich RC, Attademo AM, Zenklusen MC, Basso A. Toxicity of the fungicide trifloxystrobin on tadpoles and its effect on fish-tadpole interaction. Chemosphere. 2012;87:1348–54.
Katagiri M, Uesugi Y. Similarities between fungicidal action of isoprothiolane and organophosphorus thiolate fungicides. Phytopathology. 1977;67:1415–7.
Knoester M, Pieterse CMJ, Bol JF, Van Loon LC. Systemic resistance in *Arabidopsis* induced by rhizobacteria requires ethylene-dependent signalling at the site of application. Mol Plant-Microbe Interact. 1999;12:720–7.
Kodama O, Yamada H, Akatsuka T. Kitazin P, inhibitor of phosphatidylcholine biosynthesis in *Pyricularia oryzae*. Agric Biol Chem. 1979;43:1719–25.
Kuck KH, Stenzel K, Vors JP. Sterol biosynthesis inhibitors. In: Kramer W, Schirmer U, Jeschke P, Witschel M, editors. Modern crop protection compounds. 2nd ed. Weinheim: Wiley-VCH; 2012. p. 761–806.
Kumar PMK, Veerabhadraswamy AL. Appraise a combination of fungicides against blast and sheath blight diseases of paddy (*Oryza sativa* L.). J Exp Biol Agril Sci. 2014;2(1):49–57.
Kurahashi Y, Hattori TLT, Kagabu S, Pontzen R. Mode of action of the novel rice blast fungicide KTU 3616. Pestic Sci. 1996;47(2):199–200.
Kurahashi Y, Sakawa S, Kimboraund T, Kagabu S. Biological activity of carpropamid (KTU 3616): a new fungicide for rice blast disease. J Pestic Sci. 1997;22:108–12.
Liu L, Jiang C, Wu ZQ, Gong YX, Wang GX. Toxic effects of three strobilurins (trifloxystrobin, azoxystrobin and kresoxim-methyl) on mRNA expression and antioxidant enzymes in grass carp (*Ctenopharyngodon idella*) juveniles. Ecotoxicol Environ Saf. 2013;98:297–302.
Manandhar HK. Seed treatment against rice leaf blast. Nepal J Agric Sci. 1984;15:189.
Manandhar HK, Thapa BJ, Amatya P. Efficacy of various fungicides on the control of rice blast disease. J Inst Agric Anim Sci. 1985;6:21–9.
Matsuura K, Ishida Y, Kuragano T, Konishi K. Development of a new fungicide, ferimzone. J. Pestic. Sci. 1994;19:325–327.
Mariappan V, Rajeswari E, Kamalakannan A. Management of rice blast, *Pyricularia oryzae* by using neem (*Azadirachta indica*) and other plant products. In: Mariappan V, editor. Neem for the management of crop diseases. New Delhi: Associated Publishing Co.; 1995. p. 3–10.
Mashiko M, Kataoka T, Ichisei M, Shinkawa M. Development of a fungicide for rice: metominostrobin. J Pestic Sci. 2001;26(2):203–14.
Mbodi Y, Gaye S, Diaw S. The role of tricyclazole in rice protection against blast and cultivar improvement. Parasitica. 1987;43:187–98.
Minami E, Ando I. Analysis of blast disease resistance induced by probenazole in rice. J Pestic Sci. 1994;19:79–83.
Misato T, Ishii I, Asakawa M, Okimoto Y, Fukunaga K. Antibiotics as protectant fungicides against rice blast. Jpn J Phytopathol. 1959;24(5):302–6.
Miura H, Ito H, Takahashi S. Occurrence of resistant strains of *Pyricularia oryzae* to kasugamycin as a cause of the diminished fungicide activity to rice blast. Ann Phytopathol Soc Jpn. 1975;41:415–7.
Mizutani A, Miki N, Yukioka H, Tamura H, Masuko M. A possible mechanism of control of rice blast disease by a novel alkoxyiminoacetamide fungicide, SSF126. Phytopathology. 1996;86:295–300.
Moletti M, Giudici ML, Nipoti E, Villa B. Chemical control trials against rice blast in Italy. Inf Fitopatol. 1988;38:41–7.

Morton V, Staub T. A short history of fungicides. Online. APSnet Features. 2008. https://doi.org/10.1094/APSnetFeature-2008-0308.

Mueller DS. "Fungicides: triazoles". Integrated Crop Management News. 2006. p. 1274. http://lib.dr.iastate.edu/cropnews/1274.

Naidu VD, Reddy GV. Control of blast (BI) in main field and nursery with some new fungicides. R.P.P. 1989;69:209.

Nakashita H, Yoshioka K, Yasuda M, Nitta T, Arai Y, Yoshida S, Yamaguchi I. Probenazole induces systemic acquired resistance in tobacco thorough salicylic acid accumulation. Physiol Mol Plant Pathol. 2002;61:197–203.

Okuno T, Furusawa I, Matsuura K, Shishiyama J. Mode of action of ferimzone (TF-164), a novel systemic fungicide for rice diseases: effect on the general metabolism of *Pyricularia oryzae*. Ann Phytopathol Soc Jpn. 1989;55:281–9.

Ou SH. Rice diseases: International Rice Research Institute. Manila, Philippines; 1985.

Pal AK, Gajjar DU, Vasavada AR. DOPA and DHN pathway orchestrate melanin synthesis in *Aspergillus* species. Med Mycol. 2013;52(1):10–8.

Pieterse CMJ, Van der Does D, Zamioudis C, Leon-Reyes A, Van Wees SCM. Hormonal modulation of plant immunity. Annu Rev Cell Dev Biol. 2012;28:489–521.

Prabhu AS, Filippi MC. Seed treatment with pyroquilon for the control of leaf blast in Brazilian upland rice. Int J Pest Manag. 1993;39(3):347–53.

Pramesh D, Muniraju KM, Guruprasad GS, Mahantashivayogayya K, Reddy BGM, Gowdar SB, Chethana BS. Bio-efficacy of a combination fungicide against blast and sheath blight diseases of paddy. J Exp Agril Int. 2016a:1–8.

Pramesh D, Nataraj K, Guruprasad GS, Mahantashivayogayya K, Reddy BGM. Evaluation of a new strobilurin group of fungicide for the management of blast disease of paddy. J Exp Agric Int. 2016b:1–6.

Raji P, Louis V. Isoprothiolane—a new fungicide for the management of rice blast. ORYZA Int J Rice. 2007;44(4):366–7.

Sah DN, Karki PB. Efficacy of seed treatment and crop management in controlling leaf blast disease of rice. In: Rice Technical Working Group Meeting held at Parwanipur, Nepal. 1988. p. 6–8.

Sakuma H, Araki Y, Tanaka K, Kinbara T, Imanishi K, Shigyo T, Kuchii Y, Ogawa M, Ishikawa R, Sawada H. Studies on a Novel Fungicide Isotianil (Routine®)–1; Biological Properties on Rice Blast and Several Diseases. Jpn J Phytopathol. 2008;74:267.

Sauter H, Ammermann E, Benoit R, Brand S, Gold RE, Grammenos W, Koehle H, Lorenz G, Muller B, Rohl F, Schirmer U, Speakman JB, Wenderoth B, Wingert H. Mitochondrial respiration as a target for antifungals: lessons learned from research on strobilurins. In: Dixon GK, Copping LG, Hollomon DW, editors. Antifungal agents-discovery and mode of action. Oxford: BIOS; 1995. p. 173–91.

Shen YF, Liu L, Gong YX, Zhu B, Liu GL, Wang GX. Potential toxic effect of trifloxystrobin on cellular microstructure, mRNA expression and antioxidant enzymes in *Chlorella vulgaris*. Environ Toxicol Pharmacol. 2014;37:1040–7.

Singh S, Mohan C, Pannu PS. Bio-efficacy of different fungicides in managing blast of rice caused by *Pyricularia grisea*. Plant Dis Res. 2014;29(1):16–20.

Singh HS, Kaushik SS, Singh CM, Negi RS. Efficacy of different fungicides against rice blast caused by *Pyricularia oryzae* (Cav.) under Field Condition in Satna District of Madhya Pradesh. Int J Curr Microbiol App Sci. 2019;8(6):63–9.

Suzuki F, Yamaguchi J, Koba A, Nakajima T, Arai M. Changes in fungicide resistance frequency and population structure of *Pyricularia oryzae* after discontinuance of MBI-D fungicides. Plant Dis. 2010;94:329–34.

Taninaka K, Kurono H, Hara T, Murata K. Rice blast controlling activities of bis (alkoxycarbonyl) ketene dithioacetals and their related compounds. J Pestic Sci. 1976;1(2):115–22.

Tokousbalides MC, Sisler HD. Effect of tricyclazole on growth and secondary metabolism in *Pyricularia oryzae*. Pestic Biochem Physiol. 1978;8:26–32.

Tokuda T, Nishiki M, Hoshi H, Shinoda K, Ishida M, Misato T. Metabolic fate of fthalide (4,5,6,7-tetrachlorophthalide) in compost. J Pestic Sci. 1976;1:283–94.
Tsubata K, Kuroda K, Yamamoto Y, Yasokawa N. Development of a novel plant activator for rice diseases, tiadinil. J Pestic Sci. 2006;31(2):174–81.
Uesugi Y. Fungal choline biosynthesis—a target for controlling rice blast. Pestic Outlook. 2001;12(1):26–7.
Umetani K, Hakuno F, Nishimura A, Yamamoto Y, Hirooka T. Studies on NNF-9850 (Tiadinil, V-GET), a novel fungicide. Abst. PSJ Ann. Meeting. 2003 (in Japanese).
Umezawa H, Hamada M, Suhara Y, Hashimoto T, Ikekawa T. Kasugamycin, a new antibiotic. Antimicrob Agents Chemother. 1965;5:753–7.
Watanabe T, Igarashi H, Matsumoto K, Seki S, Mase S, Sekizawa Y. The characteristics of probenazole (Oryzemate) for the control of rice blast. J Pestic Sci. 1977;2:291–6.
Wheeler MH, Bell AA. Melanins and their importance in pathogenic fungi. Curr Top Med Mycol. 1998;2:338–87.
Woloshuk CP, Sisler HD, Tokousbalides MC, Dutky SR. Melanin biosynthesis in *Pyricularia oryzae*: site of tricyclazole inhibition and pathogenicity of melanin-deficient mutants. Pestic Biochem Physiol. 1980;14:256–64.
Woloshuk CP, Sisler HD, Vigil EL. Action of the antipenetrant, tricyclazole, on appressoria of *Pyricularia oryzae*. Physiol Plant Pathol. 1983;22:245–59.
Yamaguchi T. Development of rice blast control techniques in Japan. Jpn Pestic Inform. 1974;18:5–9.
Yamaguchi I. Activators for systemic acquired resistance. In: Fungicidal activity chemical and biological approaches to plant protection. 1998. p. 193–219.
Yamaguchi I. Overview on the chemical control of rice blast disease. In: Rice blast: interaction with rice and control. Dordrecht: Springer; 2004. p. 1–13.
Yamaguchi I, Kubo Y. Target sites of melanin biosynthesis inhibitors. Target sites of fungicide action. 1992. p. 101–18.
Yamaguchi I, Sekido S, Misato T. The effect of non-fungicidal anti-blast chemicals on the melanin biosynthesis and infection by *Pyricularia oryzae*. J Pestic Sci. 1982;7:523–9.
Yoshioka K, Nakashita H, Klessig D, Yamaguchi I. Probenazole induces systemic acquired resistance in Arabidopsis with a novel type of action. Plant J. 2001;25:149–57.
Yukioka M, Hatayama T, Morisawa S. Affinity labelling of the ribonucleic acid component adjacent to the peptidyl recognition center of peptidyl transferase in *Escherichia coli* ribosomes. Biochim Biophys Acta. 1975;390:192–208.

Chapter 6
Wheat Blast Management: Prospects and Retrospect

B. N. Devanna, P. K. Singh, C. Parameswaran, S. Samantaray, J. L. Katara, and A. Kumar

Contents

6.1 Introduction

Wheat blast disease is caused by host-specialized hemi-biotrophic, ascomycete fungi, *Magnaporthe oryzae* Triticum (MoT) pathotype. This is one of the lethal and emerging diseases of wheat-growing areas outside the South American continent. The latest wheat blast epidemic was reported from Bangladesh during the 2016 wheat-growing season. MoT-infected wheat spikes show a typical bleached head symptom on panicles, dark grey spots on leaves and in some cases typical spindle-shaped lesions on leaves. The environmental conditions of 18–30 °C temperature and a RH of >80% during ear emergence are very optimum for its rapid outbreak.

B. N. Devanna (✉) · C. Parameswaran · S. Samantaray · J. L. Katara · A. Kumar
ICAR-National Rice Research Institute, Cuttack, Odisha, India

P. K. Singh
National Agri-Biotechnology Institute, Mohali, Punjab, India

S. C. Nayaka et al. (eds.), *Blast Disease of Cereal Crops*, Fungal Biology,
https://doi.org/10.1007/978-3-030-60585-8_6

6.2 Discovery and Spread of Wheat Blast Disease

Rice and wheat are the two largest grown cereal crops in the world. Blast disease is very commonly found in all rice- and some wheat-growing regions of the world. The reports of major outbreak of rice blast disease date back to early 1900s (Sharma et al. 2012). Though the initial report on wheat blast disease indicates its possible presence during late 1930s (Puttemans 1936), the first outbreak of MoT was reported only during 1980s (Igarashi et al. 1986). Wheat offers around 20% calories and protein to worldwide population (CSISA 2014) and blast disease is one of the most destructive diseases affecting its production and productivity. MoT leads up to 100% loss in wheat production (Igarashi et al. 1986; Urashima et al. 2009; Kohli et al. 2011). Officially the first outbreak of MoT was reported in Parana state of Brazil in 1985. Though wheat blast was initially confined to Brazil, it later spread to other Latin American countries (Duveiller et al. 2016). The first blast epidemic in Bolivia was reported in 1996, and later spread to Paraguay (2002) and Argentina (2007), where it resulted in yield losses up to 80% (Alberione et al. 2008; Viedma and Morel 2002). An estimated area of 3 million ha was affected in these Latin American countries (Cruz and Valent 2017). In the subsequent years, scientists had predicted the possible spread of this disease to other wheat-growing regions of the world. Consequently, the incidences of wheat blast disease outside Latin America were reported from Kentucky (North America) and Bangladesh (Asia) during 2011 and 2016, respectively (Fig. 6.1; Callaway 2016; Islam et al. 2016; Farman et al. 2017). However, subsequent studies revealed that the possible source of wheat blast in the USA was a host jump event from a grass, whereas the MoT strain identified from Bangladesh was found to have direct lineage with wheat blast strains from Latin America (Islam et al. 2016; Farman et al. 2017). The large-scale import of Brazilian wheat grains by Bangladesh between 2008 and 2015 could be the possible source of MoT outbreak in the latter country. During 2016 wheat-growing season, wheat blast disease was also reported from West Bengal state of India, which borders the blast-affected districts of Bangladesh (Government of India 2016; Press

Fig. 6.1 Pattern of distribution and spread of wheat blast disease across the world

Trust of India 2017; Devanna and Sharma 2018). Now wheat blast disease is well established in South American countries as well as in South Asia (Bangladesh). Therefore, there are plenty of chances that this pathogen could spread to other South Asian countries, such as India and Pakistan. Also, spread of MoT to other wheat-growing regions across the world could not be ruled out.

6.3 Genomics of Wheat Blast Pathogen

Wheat blast pathogen, MoT, is an adopted lineage of rice blast pathogen *M. oryzae*. Recently, the genome sequence of highly virulent strain of MoT, B71, has been published with a final sequence assembly of ~44.46 Mb coding for 12,141 genes (Peng et al. 2019). The RNAseq analysis of MoT transcriptome also identified 2891 upregulated and 2429 downregulated genes under *in planta* B71 samples in comparison to in vitro samples. Among them, 153 and 335 genes were specific to *in culture* and *in planta* growth conditions, respectively (Peng et al. 2019). This study further identified 174 putative effector genes, including five homologues with *M. oryzae* effectors. It has been reported that the dispensable mini-chromosomes having large transposable elements might probably have a role in enhanced virulence of MoT strains in Brazil and latest in Bangladesh, leading to the outbreak of blast disease. The genome information of MoT could be further used for finding the highly virulent effector strains, and also for sequencing other strains of MoT reported from different regions. Along with this, the availability of wheat genome sequence can be used for better understanding the molecular aspects of wheat-*Magnaporthe* pathosystem. In rice-*Magnaporthe*, the availability of genomic resources has drastically increased the speed of identification and cloning of resistance (*R*) genes and their cognate avirulent (*Avr*) genes for effective management of blast disease (Sharma et al. 2012; Devanna et al. 2014; Cruz and Valent 2017; Ray et al. 2016).

Therefore, there is further need to generate genomic resources for different MoT strains. Also, existing genomics information of rice blast pathogen *M. oryzae* can also be used for comparative genomics studies to get new information. Using this strategy, researchers have identified the homologues of rice blast effector genes: *PWT3*, *PWL4*, *AVR-Pi54*, *AVR-Pik*, *AVR-Pik-chr3*, *BAS2*, *BAS1-chr1*, *BAS1*, *BAS3*, *BAS4*, *AVR1-CO39*, *AVR-Pi9*, *AVRPiz-t*, *AVR-Pii-chr3*, *AVR-Pib*, *PWL2*, *AVR-Pii-scaf1* and *AVR-Pib* (Peng et al. 2019). Further, availability of wheat genome sequence has also facilitated the identification of MoT resistance genes.

6.4 Management Strategies

Management of wheat blast is more difficult in comparison to rice blast. This is because, unlike *M. oryzae* of rice, wheat blast pathogen follows a mixed reproductive system. In this, the sexual recombination stage is subsequently followed by asexual propagation. Therefore, this mixed reproduction system facilitates a higher

Table 6.1 Blast resistance genes reported in wheat and their *Avr* counterparts from MoT

Gene	Resistance towards	Source cultivar	Chromosomal location	References	*Avr* gene
Rmg1	*Avena* isolate of *Magnaporthe*	Norin 4	–	Takabayashi et al. (2002)	–
Rmg4	Crabgrass isolate of *Magnaporthe*	P168, Shin-chunaga, Norin 4, Norin 26, Norin 29	4A	Nga et al. (2009)	–
Rmg5	Crabgrass isolate of *Magnaporthe*	Red Egyptian and Salmon	6D	Nga et al. (2009)	–
Rmg6	*Lolium* isolate of *Magnaporthe*	Chinese Spring, Shin-chunaga and Norin 4	1D	Vy et al. (2014)	–
Rmg2	Wheat isolate of *Magnaporthe*	Thather	7A	Zhan et al. (2008)	–
Rmg3	Wheat isolate of *Magnaporthe*	Thatcher	6B	Zhan et al. (2008)	–
Rmg7	Wheat isolate of *Magnaporthe*	St24, St17, St25	2A	Tagle et al. (2015)	*AVR-Rmg8* (Anh et al. 2018)
Rmg8	Wheat isolate of *Magnaporthe*	S-615	2B	Anh et al. (2015)	*AVR-Rmg8* (Anh et al. 2018)

evolutionary potential in MoT, due to both recombinational and repeat element-mediated genomic rearrangements (Zeigler 1998; Maciel et al. 2014). Therefore, effective management strategy for wheat blast disease largely rests on the identification of durable resistance genes. Till date, no wheat line with complete resistance to MoT has been reported. Similar to the rice-*Magnaporthe* pathosystem, wheat-MoT interaction also displays typical gene-for-gene interaction (Anh et al. 2015). Also, like rice blast, both qualitative and quantitative resistance is found in wheat for blast (Cruz and Valent 2017). However, the number of resistance (*R*) and avirulence (*Avr*) genes identified and cloned is comparatively more in rice (28 and 11, respectively) (Devanna et al. 2014; Ma et al. 2015; Chen et al. 2015) than in wheat; see Table 6.1. This is largely due to higher virulence nature of MoT, and limited presence of natural resistance in wheat germplasm. However, some success has been achieved with the deployment of 2NS, a genomic translocation region from *Aegilops ventricosa* in the genetic background of some wheat lines in CIMMYT (Cruz et al. 2016).

6.5 Future Prospects

The lack of complete resistance against MoT makes its management a challenging task. Specifically, under the changing global climatic conditions, this disease could spread more easily to other wheat-growing regions than earlier expected.

The outbreak of wheat blast in Bangladesh is the clear reminder for this. Hence, there is an urgent need for exploration of wheat germplasm, from different wheat-growing regions of the world. India, for example, has a vast collection of wheat germplasm, and there could be a possible source for blast resistance. Further, the resistance response provided by 2NS genomic region shall be utilized wherever possible to limit the spread of blast disease to other wheat-growing regions of the world. The availability of recent genome editing tool like CRISPR/Cas9 and other related systems would have a potential role in devising more effective strategies for the management of MoT in wheat. Besides, availability of comparatively larger sources of resistance to blast pathogens, like rice blast, shall also be studied for the presence of race-independent resistance against wheat blast pathogen.

References

Alberione E, Bainotti C, Cettour I, Salines J. Evaluation of diseases in summer planting wheat in the Argentine-Campana NEA 2007/2008. In: Proceedings of the 7th National Congress of Wheat, Santa Rosa, La Pampa, Argentina. 2008.

Anh VL, Anh NT, Tagle AG, Vy TTP, Inoue Y, Takumi S, Chuma I, Tosa Y. *Rmg8*, a new gene for resistance to Triticum isolates of *Pyricularia oryzae* in hexaploid wheat. Phytopathology. 2015;105:1568–72.

Anh VL, Inoue Y, Asuke S, Vy TTP, Anh NT, Wang S, Chuma I, Tosa Y. Rmg8 and Rmg7, wheat genes for resistance to the wheat blast fungus, recognize the same avirulence gene *AVR-Rmg8*. 2018;19(5):1252–6. https://doi.org/10.1111/mpp.12609. Epub 2017 Dec 10.

Callaway E. Devastating wheat fungus appears in Asia for first time. Nature. 2016;532:421–2. https://doi.org/10.1038/532421a.

Chen J, Peng P, Tian J, He Y, Zhang L, Liu Z. Pike, a rice blast resistance allele consisting of two adjacent NBS–LRR genes, was identified as a novel allele at the *Pik* locus. Mol Breed. 2015;35:1–15. https://doi.org/10.1007/s11032-015-0305-6.

Cruz CD, Valent B. Wheat blast disease: danger on the move. Trop Plant Pathol. 2017;42:210. https://doi.org/10.1007/s40858-017-0159-z.

Cruz CD, Bockus WW, Stack JP, Valent B, Maciel JN, Peterson GL. A standardized inoculation protocol to test wheat cultivars for reaction to head blast caused by *Magnaporthe oryzae* (*Triticum* pathotype). Plant Health Prog. PHP. 2016;186–187. https://doi.org/10.1094/PHP-BR-16-0041.

CSISA. 2014. https://csisa.org/wp-content/uploads/sites/2/2014/06/Wheat-Blast-Priority-Brief.pdf.

Devanna BN, Sharma TR. Wheat blast disease management: cues from the advancements in molecular biology of rice-*Magnaporthe* pathosystem. J Plant Biochem Biotechnol. 2018;27:249–59. https://doi.org/10.1007/s13562-017-0435-5.

Devanna NB, Vijayan J, Sharma TR. The blast resistance gene *Pi54of* cloned from *Oryza officinalis* interacts with *Avr-Pi54* through its novel non-LRR domains. PLoS One. 2014;9(8):e104840. https://doi.org/10.1371/journal.pone.0104840.

Duveiller E, He X, Singh PK. Wheat blast: an emerging disease in South America potentially threatening wheat production. In: Bonjean AP, Angus WJ, van Ginkel M, editors. World wheat book: a history of wheat breeding 3. Paris: Lavoisier; 2016. p. 1107–22.

Farman M, Peterson G, Chen L, et al. The Lolium pathotype of *Magnaporthe oryzae* recovered from a single blasted wheat plant in the United States. Plant Dis. 2017;101(5):684–92. https://doi.org/10.1094/PDIS-05-16-0700-RE.

Government of India. Minutes of the meeting on "Occurrence of blast disease on wheat" held under the Chairmanship of Agriculture Commissioner on 28th September, 2016 at Kolkata. File no. 4–2/20 13-NFSM (Ministry of Agriculture & Farmers Welfare Department of Agriculture Cooperation & Farmers Welfare Crops Division NFSM Cell, editor). Krishi Bhawan, New Delhi. 2016. p. 14.

Igarashi S, Utiamada CM, Igarashi LC, Kazuma AH, Lopes RS. *Pyricularia emtrigo*. 1. Ocorrência de Pyricularia sp. No estado do Parana. Fitopatol Bras. 1986;11:351–2.

Islam MT, Croll D, Gladieux P, Soanes DM, Persoons A, Bhattacharjee P, Hossain MS, Gupta DR, Rahman MM, Mahboob MG. Emergence of wheat blast in Bangladesh was caused by a South American lineage of Magnaporthe oryzae. BMC Biol. 2016;14:84. https://doi.org/10.1186/s12915-016-0309-7.

Kohli MM, Mehta YR, Guzman E, De Viedma L, Cubilla LE. *Pyricularia* blast-a threat to wheat cultivation. Czech J Genet Plant Breed. 2011;47:S130–4.

Ma J, Lei C, Xu X, Hao K, Wang J, Cheng Z, et al. Pi64, encoding a novel CC-NBS-LRR protein, confers resistance to leaf and neck blast in rice. Mol Plant Microbe Interact. 2015;28(5):558–68. https://doi.org/10.1094/mpmi-11-14-0367-r.

Maciel JLN, Ceresini PC, Castroagudin VL, Zala M, Kema GHJ, Mcdonald BA. Population structure and pathotype diversity of the wheat blast pathogen *Magnaporthe oryzae* 25 years after its emergence in Brazil. Phytopathology. 2014;104:95–107.

Nga NTT, Hau VTB, Tosa Y. Identification of genes for resistance to a Digitaria isolate of *Magnaporthe grisea* in common wheat cultivars. Genome. 2009;52:801–9.

Peng Z, Oliveira-Garcia E, Lin G, Hu Y, Dalby M, Migeon P, et al. Effector gene reshuffling involves dispensable mini-chromosomes in the wheat blast fungus. PLoS Genet. 2019;15(9):e1008272. https://doi.org/10.1371/journal.pgen.1008272.

Press Trust of India. Deadly wheat blast disease spreads to Bengal districts. In: Joshi V, editor. The new Indian express. Chennai: Express Publications (Madurai) Limited; 2017.

Puttemans A. Alguns dados para servir à historia da phytopathologia no Brasil e às primeiras notificações de doenças de vegetaes neste paiz. Rodriguesia. 1936;2:17–36.

Ray S, Singh PK, Gupta DK, Mahato AK, Sarkar C, Rathour R, Singh NK, Sharma TR. Analysis of *Magnaporthe oryzae* genome reveals a fungal effector, which is able to induce resistance response in transgenic rice line containing resistance gene, *Pi54*. Front Plant Sci. 2016;7:1140. https://doi.org/10.3389/fpls.2016.01140.

Sharma TR, Rai AK, Gupta SK, Vijayan J, Devanna BN, Ray S. Rice blast management through host-plant resistance: retrospect and prospects. Agric Res. 2012;1:37. https://doi.org/10.1007/s40003-011-0003-5.

Tagle AG, Chuma I, Tosa Y. *Rmg7*, a new gene for resistance to Triticum isolates of Pyricularia oryzae identified in tetraploid wheat. Phytopathology. 2015;105(4):495–9.

Takabayashi N, Tosa Y, Oh HS, Mayama S. A gene-for-gene relationship underlying the species-specific parasitism of Avena/Triticum isolates of *Magnaporthe grisea* on wheat cultivars. Phytopathology. 2002;92:1182–8.

Urashima AS, Grosso C, Stabili A, Freitas E, Silva C, Netto D, Franco I, Bottan JM. Effect of Magnaporthe grisea on seed germination, yield and quality of wheat. In: Wang G-L, Valent B, editors. Advances in genetics, genomics and control of rice blast disease. Dordrecht: Springer; 2009. p. 267–77.

Viedma LQ, Morel W. Añublo o Pyricularia del Trigo. Díptico. MAG/DIA/CRIA. Programa de Investigación de Trigo, CRIA, Capitán Miranda, Itapúa. 2002.

Vy TTP, Hyon G, Nga NTT, et al. Genetic analysis of host-pathogen incompatibility between Lolium isolates of *Pyricularia oryzae* and wheat. J Genet Plant Pathol. 2014;80:59–65.

Zeigler RS. Recombination in *Magnaporthe grisea*. Annu Rev Phytopathol. 1998;36:249–75.

Zhan SW, Mayama S, Tosa Y. Identification of two genes for resistance to Triticum isolates of *Magnaporthe oryzae* in wheat. Genome. 2008;51(3):216–21. https://doi.org/10.1139/G07-094.

Chapter 7
Recent Insights in Rice Blast Disease Resistance

Susheel Kumar Sharma, Devender Sharma, Ram Prasnna Meena, Manoj Kumar Yadav, Rajashekara Hosahatti, Abhishek Kumar Dubey, Parul Sharma, Sudhir Kumar, D. Pramesh, Sajad Un Nabi, S. Bhuvaneshwari, Y. Rupert Anand, Saurabh Kumar Dubey, and Thokchom Surjit Singh

Contents

S. K. Sharma (✉) · S. Bhuvaneshwari · T. S. Singh
ICAR Research Complex for NEH Region, Manipur Centre, Lamphelpat, Imphal, India

D. Sharma
Crop Improvement Division, ICAR-Vivekanada Parvatiya Krishi Anusandhan Sansthan, Almora, Uttarakhand, India

R. P. Meena
ICAR-Directorate of Medicinal and Aromatic Plants Research, Boriavi, Anand, Gujarat, India

M. K. Yadav
Crop Protection Division, ICAR-National Rice Research Institute, Cuttack, Odisha, India

R. Hosahatti
Crop Protection Division, ICAR-Vivekanada Parvatiya Krishi Anusandhan Sansthan, Almora, Uttarakhand, India

S. C. Nayaka et al. (eds.), *Blast Disease of Cereal Crops*, Fungal Biology,
https://doi.org/10.1007/978-3-030-60585-8_7

7.1 Introduction

Rice (*Oryza sativa* L.), an important cereal crop, is being consumed as a staple food by more than half of the world's inhabitants. As a food commodity and others, 23% of calorie intake consumed daily contributed solely from rice produce (Marathi and Jena 2015). "It is expected that, by 2035, an extra 116 million tonnes of rice will be required to feed the world's burgeoning population" (Seck et al. 2012). It will compel rice production to enlarge with over 30% to fulfil the staple food demands by 2030 of the swiftly accelerating world population, with restrictions to expand cultivated land and limitations of water resources for irrigation. It is grown ubiquitously in the world besides Antarctica and bears significant monetary importance. With Continent Asia dwelling to more than a fraction of the world population, 92% of the entire cultivated rice is grown in Asian countries alone. The rice crop production is affected by several biotic factors and over 70 different diseases caused by a diverse group of pathogens, viz. fungi, bacteria, viruses, and phytoplasmas (Zhang et al. 2009). Several fungal pathogens invade rice crops, yet blast is one of the most destructive and devastating diseases, leading to an enormous and significant reduction in grain yields both quantitatively and qualitatively. The invasion of blast pathogen on rice begins from seedling to late vegetative/reproductive stages affecting leaves, collar, nodes, panicle neck, panicles, and even roots. The most common and diagnostic symptom includes diamond-shaped lesions on the leaves, whereas lesions on the sheaths are comparably rare (Zhu et al. 2005). "The spores produced by the fungus at the later stages of crop growing season cause collar blast and neck

A. K. Dubey
Division of Crop Research, ICAR-Research Complex for Eastern Region, ICAR Parisar, Patna, Bihar, India

P. Sharma
Department of Crop Improvement, College of Agriculture, CSK Himachal Pradesh Krishi Vishvavidyalaya, Palampur, Himachal Pradesh, India

S. Kumar
ICAR-Indian Institute of Agricultural Biotechnology (IIAB), Namkum, Ranchi, Jharkhand, India

D. Pramesh
Rice Pathology Laboratory, All India Coordinated Rice Improvement Programme, University of Agricultural Sciences, Raichur, Karnataka, India

S. U. Nabi
ICAR-Central Institute of Temperate Horticulture, Srinagar, Union Territory of Jammu and Kashmir, India

Y. R. Anand
College of Agricultural Engineering and Post-Harvest Technology, Ranipool, Gangtok, Sikkim, India

S. K. Dubey
Division of Plant Pathology, ICAR-Indian Agricultural Research Institute, New Delhi, India

blast (Wang et al. 2014), leading to about 30% of yield loss" (Spence et al. 2014). The losses are determined by existing environmental conditions, growth stage of the plant at which infection occurs, and resistance level of cultivars. It occurs more often in rain-fed areas during the wet season, which are favorable environmental conditions for the growth and development. The low temperature in humid tropics and a warm and humid climate in subtropical regions increase the risk of blast epidemic, owing to its presence and survival at different environmental conditions in over 85 countries. The rice blast causes significant yield loss every year, which is quite enough to feed over 60 million people (Bevitori and Ghini 2014; Talbot 2003).

"The Blast disease is caused by hemibiotrophic fungal pathogen *Magnaporthe oryzae*, and based on the scientific and economic significance in the world, it is considered among the top ten fungal plant pathogens" (Dean et al. 2012). The blast pathogen is a haploid filamentous fungus, having a comparatively small genome of ~40 Mb which is composed of seven chromosomes (Dean et al. 2005). This ascomycetes fungus produces sexual spores (ascospores) in a structure called asci and taxonomically classified under the family *Magnaporthaceae*. The macroconidia produced on conidiophores are pyriform in shape. The conidiogenous cells of the fungus are polyblastic in nature and integrated on the conidiophores. These are sympodial, geniculate, cylindrical in shape, and denticulated. As conidia germinate, an appressorium is developed at the tip of the germ tube by which it gets attached to the surface of plant tissues, followed by an infection peg from the appressorium that penetrates plant tissues. The melanin pigments the wall of conidiophores and appressorium (Talbot 2003). The main factors that create hurdle in managing this fungus are highly variable virulence and its genetic plasticity. The infection cycle of fungus starts when conidia land on leaves of young seedlings, and produce lesions on the plant surface. The cycle ends with the dispersal of many new airborne spores after repeated sporulation for about 20 days. Under favorable environmental conditions of moisture (high humidity, extended days of plant surface wetness, night temperatures between 12 and 32 °C and slight or no wind at night), the infection cycle can last for many days (Srivastava et al. 2017). Thus, *M. oryzae* is one of the most devastating biotic threats to food security worldwide.

To combat blast disease and overcome the losses, many management strategies, *viz.* host plant resistance, biological control, chemical control, disease forecasting, and conventional breeding, have been adopted. Many of the management practices beneficial in decreasing plant diseases are having limitations in control of blast disease of rice. Since blast disease is prevailing in most rice-growing areas, and when pathogens have a wide host range, cultural practices such as eradication and crop rotation are of little value. Reduction in losses caused by this disease could be a vital factor in ensuring food security. Although many efforts have been made considering the studies on the genetics of rice, yet, continuously, it is the most destructive disease of rice. Therefore, management strategies for mitigating the yield losses are the need of the hour and should be implemented urgently considering environment friendliness, sustainability, and economic feasibility. However, to control this disease, conventional and traditional approaches are not enough due to high labor costs and more time taken (Srivastava et al. 2017).

Conventional breeding is one of the robust approaches but has some shortcomings such as linkage drag, where undesired character closely linked with resistance gene also gets transferred. Exploration of the host resistance is one of the most effective and economically viable methods for disease management. But in the case of rice blast, success is short-lived due to the occurrence of lineages (that may comprise different physiologic races) overcoming host resistance. Earlier studies on host resistance were mostly confined to the nature of resistance (Hayashi et al. 2006). The advanced molecular breeding methods, particularly marker-assisted selection (MAS), are robust, highly efficient, simpler, and precise as compared to the conventional breeding for disease resistance which is time consuming, labor intensive, and highly dependent on environmental conditions (Nguyen et al. 2006). Besides, in traditional breeding identification of promising trait depends upon the phenotype of plants. Accomplishment of field resistance is a long process for screening with the regular occurrence of new virulent races of the pathogen; thus development of newly improved varieties with the resistance cannot be guaranteed (Zhang et al. 2006). Hence, the MAS is a potent tool in blast resistance breeding because resistance phenotypes are encoded by single or few genes and trapping through advantageous MAS is quite simple as the blast disease control is governed by a certain interaction of a particular resistance (R) gene with a specific avirulence gene in the pathogen. At present, molecular markers play an important role in improving the efficacy of conventional breeding by carrying out direct selection of the trait of interest and also linked molecular markers of that trait. The progress in the advancement of molecular markers and functional genomics for blast disease resistance has been strengthened continuously over time (Hayashi et al. 2006). The identified resistant lines against blast disease-carrying both major and minor R-genes are the essential genetic resource for rice breeders as well as breeding programs, by which it will improve blast resistance in elite rice varieties. Hence, identification of these R-genes/alleles aided by molecular and genomic tools will accelerate the contemporary plant breeding program with the incorporation of advanced genetic and genomic resources (Li et al. 2014).

About 145 quantitative blast R-genes have been identified in rice, so far, and 36 R-genes have been efficaciously cloned and characterized (Ashkani et al. 2016). Conventional breeding methods and MAS tools together expedite R-gene pyramiding in elite rice varieties to strengthen blast resistance and durability. The inclusive approaches of quantitative trait locus (QTL) association mapping, allele mining, bioinformatics tools, and genetic engineering have contributed to the strengthening of molecular breeding for development of blast-resistant cultivars (Coca et al. 2004). The advancement in bioinformatics tools also plays a crucial role in studying R and Avr genes in the evolved new *Magnaporthe oryzae* pathotype-rice ecosystems. Genome editing, a robust tool for the creation of variation in the gene pool, is a recent development which interrupts the functions of the gene either by inserting or by deleting the target gene. Moreover, clustered regularly interspaced short palindromic repeats/CRISPR-associated protein 9 (CRISPR/Cas9) genome editing method is more precise and uses single sequence-specific nucleases which lead to the targeted cleavage of genomic DNA (Rani et al. 2016). Such advanced technologies facilitate the frequent introgression of single as well as multiple disease resistance genes into blast-susceptible promising rice varieties. This chapter describes genetic mapping and characterization of blast resistance genes, and defense-regulator genes as novel protagonists for broad-spectrum blast resistance.

7.2 Genetics and Mapping of Resistance Genes Against Leaf Blast Disease

Characterization and exploitation of disease-resistant genes/QTLs and their further deployment in various cultivars are the most preferred approaches to breed the disease-resistant varieties. The resistant genes for rice blast fungus, *M. oryzae*, were initially reported in Japan by Sasaki (1922). The first rice blast gene Pi-a was identified from a *japonica* rice variety "Aichi Asahi" by Kiyosawa (1967). So far, about 145 R-genes for rice blast resistance have been identified and mapped from both *indica* and *japonica* subspecies of rice (Table 7.1). These R-gene donors/source cultivars of rice containing two or more R-genes have been compared (Fig. 7.1). Cultivar IR 64 harbors maximum of 11 R-genes which are localized on almost all rice chromosomes. Out of the identified R-genes, 36 have been cloned and molecularly characterized till date (Bao-hua et al. 2018).

The molecular maps for each 12 rice chromosomes helped in utilizing the available marker information for the discovery and localization of resistance genes. Two blast resistance genes Pi2 and Pi4 were firstly mapped by Yu et al. (1991) from near-isogenic lines (NILs) set by using restriction fragment length polymorphism (RFLP), which were positioned on chromosomes 6 and 12, respectively. "From the first genetic map of rice, based on RFLP markers (McCouch et al. 1988), various genetic markers including single-nucleotide polymorphisms (SNPs), small insertions/deletions (InDels), simple sequence repeats (SSRs) and amplified fragment length polymorphisms (AFLPs) have been deployed for constructing the saturated linkage maps of rice" (Monna et al. 1994; Chen et al. 1997). The whole-genome sequence information of *indica* and *japonica* subspecies of rice made it possible to scan for SNPs and InDels linked to the blast resistance genes (Liu et al. 2012; Takagi et al. 2013; Deng et al. 2017). These developments helped researchers for the discovery and localization of resistance genes and linked molecular markers on specific chromosomes.

Of the total mapped R-genes, a majority of genes (>50%) are located on chromosomes 2, 6, 11, and 12 which harbor 13, 26, 38, and 28 genes, respectively (Fig. 7.2), whereas 3 and 7 chromosomes harbor a minimum (~1%) of 2 and 1 gene, respectively. Shang et al. (2009) reported Pid3 R-gene by *in silico* comparison of paired NBS-LRR genes between *indica* (93-11) and *japonica* rice (Nipponbare) genome sequences. Approximately 500 QTLs for rice blast resistance have been identified in 15 different mapping populations of *indica* and *japonica* crosses (Ballini et al. 2008; Li et al. 2019). "Of these QTLs, 23 blast resistance loci such as PiGD-1(t), Pi25(t), Pi26(t), Pi27(t), Pitq1, Pitq5, Pitq6, Pizh, Pi24(t), Pi25(t), Pi28(t), Pi29(t),

Table 7.1 List of identified and mapped blast resistance R-genes

S. no.	Gene	Chromosome	Genomic position (Mb)	Linked marker	Source cultivar	Country	References
1	Pit	1	2.27	RFLP, SNP	K-59, Tjahaja, K-59	Japan	Kaji and Ogawa (1996) and Hayashi et al. (2006)
2	Pi27(t)	1	5.55	SSR	Q14 and Q61	France	Zhu et al. (2004)
3	Pi-h2(t)	1	7.9	SSR	HR4	India	Xiao et al. (2015)
4	Pi-tp(t)	1	25.13	SSR	CO39 and Tetep	India	Barman et al. (2004)
5	Pi35(t)	1	32.1	SSR	Hokkai 188	Japan	Nguyen et al. (2006)
6	Pi64	1	32.31	SSR, InDel	Yangmaogu	Japan	Ma et al. (2015)
7	Pi 37(t)	1	33.1	SSR	Cultivar St. No. 1	China	Chen et al. (2005) and Lin et al. (2007a)
8	Pi-sh	1	33.3	SSR	Akihikari and Shin 2	Japan	Fukuta (2004)
9	Pi24(t)	1	5.24	SSR	Azuenca	France	Nguyen et al. (2006)
10	Pir2-3(t)	2	–	SSR	IR64	Indonesia	Dwinita et al. (2008)
11	Pirf2-1(t)	2	–	SSR	*O. rufipogon*	Indonesia	Dwinita et al. (2008)
12	Pi-Da(t)	2	2.21	SSR	Dacca 6	–	Shi et al. (2012) and Lei et al. (2005)
13	Pig(t)	2	34.34	SSR	Guangchangzhan	China	Zhou et al. (2004) and Shi et al. (2012)
14	Pi-25(t)	2	34.36	QTL	IR64	France	Sallaud et al. (2003) and Nguyen et al. (2006)
15	Pi-tq5	2	34.61	RFLP	Teqing	USA	Tabien et al. (2000), Tabien et al. (2002) and Zhou et al. (2004)

(continued)

Table 7.1 (continued)

S. no.	Gene	Chromosome	Genomic position (Mb)	Linked marker	Source cultivar	Country	References
16	Pi-14(t)	2	34.94	RFLP, Isozyme	Maowangu	Japan	Pan et al. (1998) and Zhou et al. (2004)
17	Pi-16(t)	2	34.94	RFLP, Isozyme	AUS373	Japan	Pan et al. (1999) and Zhou et al. (2004)
18	Pi-d1(t)	2	34.94	SSR, RFLP	Digu	China	Chen et al. (2004)
19	Pi-y2(t)	2	35.03	SSR	Yanxian No. 1	China	Fukuta et al. (2004) and Lei et al. (2005)
20	Pi-y1(t)	2	35.03	SSR	Yanxian No. 1	China	Fukuta et al. (2004) and Lei et al. (2005)
21	Pi-b	2	35.1	SNP	Tohoku, Koshihikari	Japan	Hayashi et al. (2006)
22	Pitq2	2	–	USA	Teqing	USA	Tabien et al. (1996)
23	Pi66(t)	3	26.78	SSR	AS20-1	Australia	Liang et al. (2016)
24	Pitq3	3		–	Teqing	USA	Tabien et al. (1996)
25	Pikur 1	4	24.61	Isozyme	Kuroka	Japan	Fukuoka et al. (2009) and Goto (1970)
26	pi-21	4	19.81	RFLP, SSR	Owarihatamochi	Japan	Fukuoka and Okuno (2001), Ahn et al. (1997) and Pan et al. (1997)
27	Pias(t)	4	31.26	SSR, CAPS	Asominori	China	Endo et al. (2012)
28	Pi-45(t)	4	31.49	SSR	Moroberekan	Japan	Kim et al. (2011)
29	Pikahei-1(t)	4	31.67	SSR, SNP	Kahei	–	Xu et al. (2008a)
30	Pi-39(t)	4	32.68	SSR	Mineasahi and Chubu 111	China	Terashima et al. (2008) and Liu et al. (2007c)

(continued)

Table 7.1 (continued)

S. no.	Gene	Chromosome	Genomic position (Mb)	Linked marker	Source cultivar	Country	References
31	Pi(t)	4	2.27	–	P167	–	Causse et al. (1994)
32	Pitq4	4		USA	Teqing	USA	Tabien et al. (1996)
33	Pi26(t)	5	2.78	RFLP, RAPD	IR64	France	Sallaud et al. (2003)
34	Pi23	5	10.75	RFLP, SSR	Suweon 365	Korea	Ahn et al. (1997)
35	Pi-10(t)	5	14.52	RAPD	Tongil	India	Naqvi et al. (1995) and Wu et al. (2005)
36	Pi26(t)	5	2.07	–	Azucena/Gumei 2	France	Wu and Tanksley (1993) and Ahn et al. (1996) and Nguyen et al. (2006)
37	Pi22	6	4.89	RFLP	Suweon 365	Korea	Ahn et al. (1997), Terashima et al. (2008)
38	Pi27(t)	6	6.92	RFLP	IR64	France	Sallaud et al. (2003)
39	Pi-40(t)	6	9.86	STS, SSR	IR65482, Co39, and *O. australiensis* (W)	Philippines	Jeung et al. (2007)
40	Pi2-1	6	10.08	SSR, SFP	Tianjingyeshengdao	China	Wang et al. (2012) and Qu et al. (2006)
41	Pi2-2	6	10.2	SSR	Jefferson	–	Jiang et al. (2012) and Ballini et al. (2008)
42	Pigm(t)	6	10.36	CAPS, InDel	Gumei 4	China	Deng et al. (2006)
43	Pi-9(t)	6	10.38	STS	IR31917	Philippines	Qu et al. (2006)
44	Pi51(t)	6	10.38	InDel, SSR	D69	–	Xiao et al. (2012)
45	Pi2	6	10.39	SSR, STS, RFLP	5173, C101A51	–	Jiang and Wang (2002) and Zhou et al. (2006)

(continued)

Table 7.1 (continued)

S. no.	Gene	Chromosome	Genomic position (Mb)	Linked marker	Source cultivar	Country	References
46	Piz	6	10.39	STS	Zenith, Fukunishiki, Toride 1, Tadukan	Japan	Goto (1976) and Zhou et al. (2006)
47	Piz-t	6	10.39	STS	Toride No. 1	Japan	Zhou et al. (2006)
48	Pi50(t)	6	10.41	SSR, CAPS	EBZ, EBZ × LTH F2 and (EBZ × LTH) × LTH, BC1F2	–	Zhu et al. (2012) and Jiang et al. (2012)
49	Pi59(t)	6	10.82	SSR	Haoru_US-2	Myanmar	Koide et al. (2013) and Zhou et al. (2006)
50	Pi26(t)	6	11.06	RFLP, SSR	Gumei 2	China	Wu et al. (2005)
51	Pi8	6	11.36	Isozyme markers, RFLP	Kasalath	Japan	Pan et al. (1996a) and Takehisa et al. (2009)
52	Pi-25(t)	6	12.33	RFLP, RGA, SSR	Gumei 2	China	Wu et al. (2013) and Zhuang et al. (2001)
53	Pi25	6	18.09	–	Gumei 2	China	Zhuang et al. (2001)
54	Pid3	6	13.05	STS	Digu	China	Shang et al. (2009)
55	Pi3(t)	6	–	–	Pai-kan-tao	Philippines	Mackill and Bonman (1992)
56	Pi-13	6	15.83	SSR	*O. minuta* (W), Kasalath	Philippines	Ebitani et al. (2011) and Amante-Bordeos et al. (1992)
57	Pi-dt(2)	6	17.16	SSR, RGA	Digu	China	Chen et al. (2004)
58	Pi9	6	10.39	–	Cultivar TP309	–	Qu et al. (2006) and Koide et al. (2013)
59	Pid2	6	17.16	CAPS	Digu	China	Chen et al. (2006)
60	Pii1	6	2.29	–	Fujisaka 5	Japan	Pan et al. (1996b)

(continued)

Table 7.1 (continued)

S. no.	Gene	Chromosome	Genomic position (Mb)	Linked marker	Source cultivar	Country	References
61	Pi-tq1	6	29.02	RFLP	Teqing	USA	Tabien et al. (2000)
62	Piz-5	6	–	–	C101A51_CO39	–	Deng et al. (2006)
63	Pi-17(t)	7	22.25	Isozyme marker	DJ 123	Philippines	Pan et al. (1996b) and Zhu et al. (2012)
64	Pi-11(t)	8	13.93	RFLP, RAPD	Zhai-Ye-Quing	China	Causse et al. (1994)
65	Pi-33	8	7.56	SSR, RFLP	IR64, Bala	France	Berruyer et al. (2003) and Sallaud et al. (2003)
66	Pi-29(t)	8	13.93	RFLP, RAPD, Isozyme	IR64	France	Sallaud et al. (2003) and Nguyen et al. (2006)
67	PiGD-1(t)	8	16.37	SSR, RFLP, RGA	Sanhuangzhan 2	China	Liu et al. (2004) and He et al. (2012)
68	Pi-36(t)	8	2.87	SSR, CRG	Q61	China	Liu et al. (2005)
69	pi55(t)	8	25.58	SSR, STS	Yuejingsimiao 2	China	Xiu-Ying et al. (2012)
70	Pizh	8	4.38	–	Zhai-Ya-Quing8	China	Sallaud et al. (2003)
71	Pi-5(t)	9	9.77	AFLP, RFLP, CAPS	RIL249, Moroberekan	Philippines	Jeon et al. (2003)
72	Pi56(t)	9	9.77	SSR, CRG, SNP	SHZ-2	–	Jeon et al. (2003)
73	Pihk2	9	10.17	SSR, ILP, InDel	Heikezijing	–	He et al. (2012)
74	Pii	9	2.29	–	Ishikari Shiroke	Japan	Ise (1991) and Shinoda et al. (1971)
75	Pi15	9	9.61	SSR, CRG	Q61 and GA25	China	Lin et al. (2007b) and Liu et al. (2004)

(continued)

Table 7.1 (continued)

S. no.	Gene	Chromosome	Genomic position (Mb)	Linked marker	Source cultivar	Country	References
76	Pii2	9	1.03	–	Ishikari Shiroke	Japan	Pan et al. (2003), Kinoshita and Kiyosawa (1997)
77	Pi3(t)	9	7.8	–	Pai-kan-tao	–	Kinoshita and Kiyosawa (1997)
78	PiGD-2(t)	10	–	SSR, RFLP, RGA	Sanhuangzhan 2	–	Liu et al. (2004)
79	Pi28(t)	10	21.04	RFLP, RAPD	Azucena, IR64	–	Sallaud et al. (2003)
80	Pi-30(t)	11	4.41	RFLP, RAPD, Isozyme	IR64	France	Sallaud et al. (2003) and Nguyen et al. (2006)
81	Pi-a	11	6.49	SSR, InDel	Aichi Asahi	Japan	Zeng et al. (2011)
82	Pi60(t)	11	6.62	SSR, InDel	93-11	China	Lei et al. (2013)
83	Pi-CO39(t)	11	6.66	SSR, RFLP	Co39	USA	Chauhan et al. (2002) and Huang et al. (2011)
84	Pi-7(t)	11	18.64	12.37	RIL29 (Moroberekan)	USA	Wang et al. (1994)
85	Pi-34	11	19.96	SSR	Chubu-32	Japan	Zenbayashi et al. (2007)
86	Pi-38	11	22.48	SSR, AFLP	Tadukan	India	Gowda et al. (2006)
87	Pik-h	11	24.99	SNP	IRBLkh-K3, HP2216, and Tetep	India	Xu et al. (2008b)
88	Pi54	11	25.26	SSR	Tetep	India	Sharma et al. (2005)
89	Pik-s	11	27.31	SSR	Shin 2	Japan	Fjellstrom et al. (2004)
90	Pi-jnw1	11	27.36	SSR, InDel	Jiangnanwan	–	Wang et al. (2016)
91	Pise	11	5.74	–	Sensho	Japan	Goto (1970)

(continued)

Table 7.1 (continued)

S. no.	Gene	Chromosome	Genomic position (Mb)	Linked marker	Source cultivar	Country	References
92	Pi-hk1	11	27.66	SSR	Heikezijing	–	Wu et al. (2013) and Liu et al. (2012)
93	Pi43(t)	11	27.67	SSR	Zhe733	–	Lee et al. (2009a)
94	Pi-47	11	27.67	SSR	Xiangzi 3150	China	Huang et al. (2011) and Ahn et al. (2000)
95	Pik-l	11	27.69	SSR, STS, CAPS	Liziangxintuanheigu, Kusabue	China	Singh et al. (2015), Hayasaka et al. (1995) and Hayashi et al. (2006)
96	Pi46(t)	11	27.74	SSR, InDel	H4	–	Xiao et al. (2011)
97	Pikur2	11	2.84	–	Kuroka	Japan	Goto (1988)
98	Pi-1(t)	11	28	STS, RFLP, SSR, CAPS	Apura, C101LAC	USA	Parco (1995), Yu et al. (1996), Fuentes et al. (2008) and Hua et al. (2012)
99	Pik-m	11	28	RFLP, SSR	Tohoku IL4, Tsuyuake	China	Kaji and Ogawa (1996) and Li et al. (2007)
100	Pikg	11	27.31	–	GA20	Japan	Pan et al. (1996a)
101	Pik-e	11	28	SSR, InDel	Xiangzao 143	China	Chen et al. (2015)
102	Pi-k	11	28.01	RFLP, InDel, SNP	Kusabue, Kanto 51	China	Hayasaka et al. (1996) and Hayashi et al. (2006)
103	Piis1	11	2.84	–	Imochi Shirazu	Japan	Goto (1970)
104	Pik-p	11	28.05	SSR, CAPS	K60, HR22	China	Wang et al. (2009)

(continued)

Table 7.1 (continued)

S. no.	Gene	Chromosome	Genomic position (Mb)	Linked marker	Source cultivar	Country	References
105	Pi-h1(t)	11	28.11	SSR, InDel	HR4	India	Xiao et al. (2015)
106	Pif	11	24.69	–	Chugoku 31-1	Japan	Shinoda et al. (1971)
107	Pi65(t)	11	28.22	SNP, InDel	Gangyu 129	–	Zheng et al. (2016)
108	Pi49	11	28.8	SSR	Mowanggu	–	Sun et al. (2013) and Chen et al. (1999)
109	Pib2	11	26.79	–	Lemont	Philippines	Tabien et al. (1996)
110	Pi-44(t)	11	28.93	RFLP, STS, AFLP	Moroberekan	USA	Chen et al. (1999) and Chauhan et al. (2002)
111	Pi18	11	28.93	RFLP	Suweon365	Korea	Ahn et al. (2000)
112	Pi-lm2	11	28.93	RFLP	Lemont, Teqing	USA	Tabien et al. (2000)
113	Mpiz	11	4.07	–	Zenith	Japan	Goto (1976)
114	Pb1	11	21.71	–	Modan	Japan	Fujii et al. (1999) and Hayashi et al. (2010a)
115	PBR	11	–	RFLP, SSR	St. No. 1	Japan	Fukuoka and Okuno (2001)
116	Pi1	11	26.49	RFLP	LAC23	Philippines	Yu et al. (1991)
117	Pish	11	33.38	–	Nipponbare	Japan	Imbe and Matsumoto (1985)
118	Pi-6(t)	12	7.73	RFLP	Apura	USA	McCouch et al. (1994)
119	Pi12	12	7.73	RFLP	Hong Jiao Zhan K80-R-Hang Jiao-Zhan	Japan	Zhuang et al. (1998)
120	Pi62(t)	12	7.73	RAPD, RFLP	Yashiromochi	Japan	Wu et al. (1996)
121	Pi62(t)	12	7.73	RAPD, RFLP	Yashiromochi	Japan	Wu et al. (1996)

(continued)

Table 7.1 (continued)

S. no.	Gene	Chromosome	Genomic position (Mb)	Linked marker	Source cultivar	Country	References
122	Pi-tq6	12	7.73	RFLP	Teqing	USA	Tabien et al. (2000)
123	Pitb	12	9.37	SSR, InDel	Zixuan	–	Sun et al. (2013)
124	Pita3(t)	12	9.89	SSR	IRBLta2-Re	–	Chen et al. (2015)
125	Pi61(t)	12	9.98	InDel, SSR	93-11	China	Lei et al. (2013)
126	Pi58(t)	12	10.42	SSR	Haoru	Myanmar	Koide et al. (2013)
127	Pita	12	10.6	RFLP, RAPD, SNP	Tadukan, Yashiromochi	USA	Rybka et al. (1997), Hayashi et al. (2006) and Bryan et al. (2000)
128	Pita-2	12	10.6	RFLP, RAPD, SNP	Yashiromochi, Pi No. 4	Japan	Rybka et al. (1997) and Hayashi et al. (2006)
129	Pi-24(t)	12	10.6	RFLP, RAPD, RGA	Zhong 156	–	Zhuang et al. (2002)
130	Pi-39	12	10.61	SSR	Q-15 and Chubu 111	China	Liu et al. (2007c)
131	Pi-42(t)	12	10.62	RAPD, SSR, STS	DHR9	India	Kumar et al. (2010)
132	Pi-19(t)	12	10.73	SSR	IRBL19-A and Aichi Asahi	Japan	Koide et al. (2011) and Hayashi et al. (1998)
133	Pi57(t)	12	10.8	SSR, STS	IL-E1454	Myanmar	Dong et al. (2017)
134	Pi-31(t)	12	11.93	RFLP, RAPD,	IR64	France	Sallaud et al. (2003)
135	Pi-48	12	11.95	SSR	Xiangzi 3150	China	Huang et al. (2011)
136	Pi51(t)	12	11.95	SSR, SFP	Tianjingyeshengdao	China	Wang et al. (2012)
137	Pi67	12	12.09	SSR	Tetep	India	Joshi et al. (2019)
138	Pi157	12	12.37	RFLP	Moroberekan	India	Naqvi and Chattoo (1996)

(continued)

Table 7.1 (continued)

S. no.	Gene	Chromosome	Genomic position (Mb)	Linked marker	Source cultivar	Country	References
139	Pi-20(t)	12	12.95	SSR	IR64	Philippines	Li et al. (2008) and Imbe et al. (1997)
140	Pih3(t)	12	12.95	SSR	HR4	India	Xiao et al. (2015)
141	PiGD-3(t)	12	14.45	SSR, RFLP, RGA	Sanhuangzhan 2	China	Liu et al. (2004)
142	Pi-41	12	16.74	SSR, STS	93-11	China	Yang et al. (2009)
143	Pi-32(t)	12	21.24	RFLP, RAPD	IR64	France	Sallaud et al. (2003)
144	Pi39(t)	4, 12	–	SSR	Chubu 111, Q15	China	Liu et al. (2007c)
145	Pi51(t)	12	–	–	Tianjingyeshengdao	China	Qu et al. (2006)

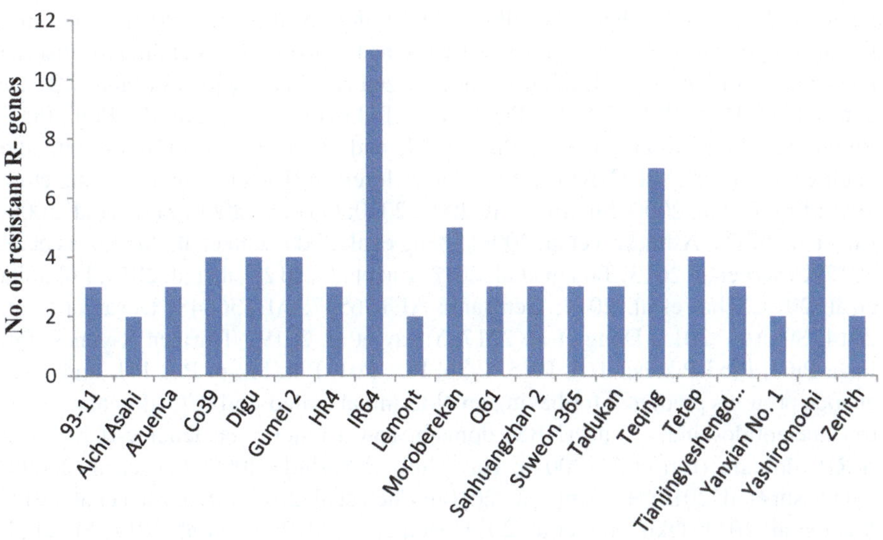

Fig. 7.1 Comparison of source cultivars/donors containing two or more than two R-genes

Pi30(t), Pi31(t), Pi32(t), PiGD-3(t), Pi35(t), PiGD-2(t), Pilm2, Pi7(t), Pi34, pi21 have been discovered within these QTLs" (Table 7.1). Some of the major genes such as Pikh, Pi-1, Pi9, Pi20, Pi27, Pi39, Pi40, and Pit impart broad-spectrum resistance (BSR) (Liu et al. 2002), and some of them including Pia, Pib, Pii, Pi-km, Pi-t, Pi12, and Pi19 confer race-specific resistance (RSR) (Yang et al. 2009).

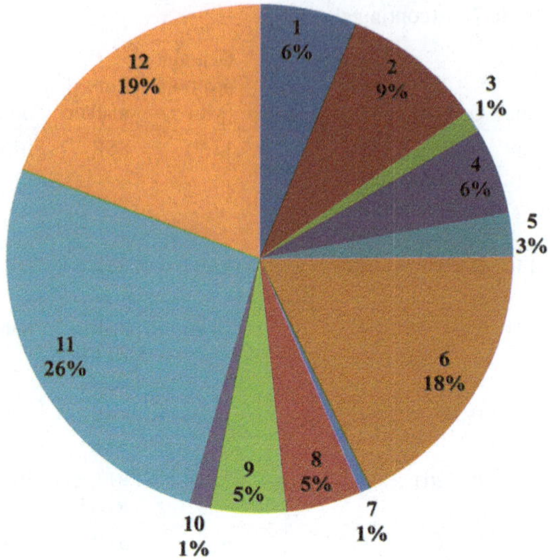

Fig. 7.2 Chromosome-wise distribution of blast resistance R-genes

Based on the fine mapping of rice blast resistance R-genes with linked molecular markers, a map-based cloning approach can be utilized for the molecular characterization of R-genes. Further, this will lead to the identification of structural and functional components of these R-genes, which is the basis for understanding the disease resistance. Among the molecularly characterized rice blast R-genes (Table 7.2), 20 genes, Pi37, Pi35, Pi-b, Pi9, Pi2, Piz-t, Pi-d3, Pid3-A4, Pi50, Pigm, Pii, Pi56, Pi54, Pikm, Pi54rh, Pi-CO39, PiK-h, Piks, Pi54, and Pi-ta encode proteins with the nucleotide-binding site (NBS) and leucine-rich repeat (LRR) domains (Wang et al. 1999; Bryan et al. 2000; Sharma et al. 2005, 2010; Qu et al. 2006; Zhou et al. 2006; Lin et al. 2007a; Ashikawa et al. 2008; Shang et al. 2009; Zhu et al. 2012; Das et al. 2012; Cesari et al. 2013; Takagi et al. 2013; Liu et al. 2012; Lü et al. 2013; Fukuoka et al. 2014; Zhai et al. 2014; GenBank: AET36547; AET36548; Devanna et al. 2014; Su et al. 2015; Deng et al. 2017; Yadav et al. 2019). Thirteen R-genes, Pit, Pish, Pi64, Pi63/Pikahei-1(t), Pi25, Pi36, Pi5, Pb1, Pik, Pik-p, Pia, Pi1, and Pike, encode proteins predicted for having an N-terminal coiled-coil (CC) domain, a central nucleotide-binding site (NBS) domain, and a C-terminal leucine-rich repeat (LRR) domain (Liu et al. 2007b; Hayashi and Yoshida 2009; Lee et al. 2009b; Takahashi et al. 2010; Hayashi et al. 2010b; Chen et al. 2011, 2015; Zhai et al. 2011; Yuan et al. 2011; Okuyama et al. 2011; Hua et al. 2012; Xu et al. 2014; Ma et al. 2015). Pi-d2 encodes unique protein B-lectin receptor kinase (Chen et al. 2006). "Whereas, pi21 R gene encodes a proline-rich heavy metal binding protein (Fukuoka et al. 2009) and Pitr encodes a putative E3 ligase" (Zhao et al. 2018).

Genetic mapping and molecular cloning of the blast resistance R-genes provide the basis for their efficient utilization in the molecular breeding programs by assisting in the selection of R-genes with the help of DNA markers (Singh et al. 2012), and pyramiding two or more than two R-genes for accomplishing broad-spectrum and durable resistance (Hittalmani et al. 2000). This also provided information on

Table 7.2 List of cloned and characterized blast resistance R-genes

S. no.	R-gene	Chromosome	Cloning strategy	Protein type	Donor	Reference
1	Pi37	1	MB	NBS–LRR	St. No. 1	Lin et al. (2007a)
2	Pit	1	MB	CC–NBS–LRR	K59	Hayashi and Yoshida (2009)
3	Pish	1	Mutant screening	CC–NBS–LRR	Nipponbare	Takahashi et al. (2010)
4	Pi35	1	MB	NBS-LRR	Hokkai 188	Fukuoka et al. (2014)
5	Pi64	1	MB	CC–NBS–LRR	Yangmaogu	Ma et al. (2015)
6	Pi-b	2	MB	NBS–LRR	Tohoku IL9	Wang et al. (1999)
7	pi21	4	MB	Proline-rich metal-binding protein	Owarihatamochi	Fukuoka et al. (2009)
8	Pi63/Pikahei-1(t)	4	MB	CC–NBS–LRR	Kahei	Xu et al. (2014)
9	Pi9	6	MB	NBS–LRR	75-1-127	Qu et al. (2006)
10	Pi2	6	MB	NBS–LRR	Jefferson	Zhou et al. (2006)
11	Piz-t	6	MB	NBS–LRR	Zenith	Zhou et al. (2006)
12	Pi-d2	6	MB	B lectin receptor kinase	Digu	Chen et al. (2006)
13	Pi-d3	6	In silico analysis	NBS–LRR	Digu	Shang et al. (2009)
14	Pi25	6	MB	CC–NBS–LRR	Gumei2	Chen et al. (2011)
15	Pid3-A4	6	MB	NBS-LRR	A4 (*Oryza rufipogon*)	Lü et al. (2013)
16	Pi50	6	MB	NBS-LRR	Er-Ba-zhan (EBZ)	Zhu et al. (2012), Su et al. (2015)
17	Pigm	6	MB	NBS–LRR	Gumei4	Deng et al. (2017)
18	Pi36	8	MB	CC–NBS–LRR	Kasalath	Liu et al. (2007b)
19	Pi5	9	MB	CC–NBS–LRR	RIL260	Lee et al. (2009b)
20	Pii	9	Mutant screening	NBS–LRR	Hitomebore	Takagi et al. (2013)

(continued)

Table 7.2 (continued)

S. no.	R-gene	Chromo-some	Cloning strategy	Protein type	Donor	Reference
21	Pi56	9	MB	NBS–LRR	Sanhuangzhan No. 2	Liu et al. (2012)
22	Pi54	11	MB	NBS–LRR	Tetep	Sharma et al. (2005)
23	Pikm	11	MB	NBS–LRR	Tsuyuake	Ashikawa et al. (2008)
24	Pb1	11	MB	CC-NBS-LRR	Modan	Hayashi et al. (2010a)
25	Pik	11	MB	CC–NBS–LRR	Kusabue	Zhai et al. (2011)
26	Pik-p	11	MB	CC–NBS–LRR	K60	Yuan et al. (2011)
27	Pia	11	MB and mutant screening	CC–NBS–LRR	Sasanishiki	Okuyama et al. (2011)
28	Pi1	11	MB	CC–NBS–LRR	C101LAC	Hua et al. (2012)
29	Pi54rh	11	MB	NBS-LRR	*Oryza rhizomatis* (nrcpb 002)	Das et al. (2012)
30	Pi-CO39	11	yeast two-hybrid screening	NBS-LRR	CO39	Cesari et al. (2013)
31	Pi54of	11	MB	NBS–LRR	*Oryza officinalis* (nrcpb004)	Devanna et al. (2014)
32	PiK-h	11	Positional cloning	NBS-LRR	K3	Zhai et al. (2014)
33	Pike	11	MB	CC–NBS–LRR	Xiangzao143	Chen et al. (2015)
34	Piks	11		NBS–LRR	Unknown	GenBank: AET36547.1, AET36548.1
35	Pi-ta	12	MB	NBS–LRR	Yashiromochi	Bryan et al. (2000)
36	Pitr	12	MB	Putative E3 ligase	Katy	Zhao et al. (2018)

an array of gene-linked, gene-based, and functional markers which can augment conventional resistance breeding programs. Marker-assisted backcross breeding (MABB) and gene pyramiding have presented an opportunity to transfer a single R-gene or a combination of two or more genes for the development of blast-resistant cultivars. "Transfer of blast resistance R genes such as Pita, Piz-5 and Pi1 into rice variety Co39 (Hittalmani et al. 2000), Pi54, Pita, Pi-b, Pi2, Pi9, Pi1 and Pi5 genes into varieties such as Pusa Basmati 1 (Khanna et al. 2015), Pi2 and Pi 54 into Pusa 1883, and Piz 5 and Pi 54 into Pusa 1884 has been successfully accomplished" (Ellur et al. 2016). Many donor lines have been developed for blast resistance through MABB such as Pusa 1602 (Pi2) and Pusa 1603 (Pi 54) and further their transfer into PB 6 and PB 1121 (Singh et al. 2012).

7.3 Structure and Functions of R-Genes

To ward off the numerous pathogens and biotic organisms in nature, plants employ the multifaceted and sophisticated immune or defense system. The immune system of plants acts in multilayered surveillance, governed by various extra- and intracellular receptor molecules. R-genes are the most effective and powerful tools against pathogen invasion as they can specifically identify the effector molecules or associated proteins of the corresponding pathogen to stimulate the plant immune reaction at the site of infection. When the pathogen invades the plant, an incompatible reaction from host-pathogen interaction occurs with programmed cell death, termed as the hypersensitive response (HR), in the vicinity of infected cells (Heath 2000). Interestingly, each form of cell death has its particular impact on inflammation and on development of innate and adaptive immune responses.

7.3.1 Classes of R-Genes

The resistance genes in plants can be broadly classified into eight groups based on their amino acid motif organization and localization of their coded protein in the cells (Table 7.3) (Gururani et al. 2012). R-gene structure mainly comprises a nucleotide-binding site (NBS), leucine-rich repeats (LRR), transmembrane domain (TM), coiled-coil region (CC), toll-interleukin-1 receptor (TIR) domain, and protein kinase domain (KIN).

In plants, NBS-LRR genes signify the largest group of R-genes which encode proteins with variable N-terminus domain of approximately 200 amino acids (aa). They are connected by NBS domain of nearly 300 aa and a variable tandem array of around 10–40 short LRR motifs. Further, on the basis of motif within their N-terminus, these NBS-LRR genes are classified into three subgroups, *viz.* TIR group, CC group, and non-motif group.

The NBS domain is also present in various proteins with ATP- or GTP-binding activity and involved in various activities like cell growth, differentiation, cytoskeletal organization, vesicle transport, apoptosis, and defense, such as ATP synthase β subunits, Ras protein, ribosomal elongation factors, and adenylate kinase (Traut 1994). The NBS domain plays an important role in defense signaling mechanism in plants. The LRR domain is responsible for R-Avr protein recognition (Jones and Jones 1997).

The TIR domain shares homology with the intracellular regions of the Drosophila protein TOL1. Another coiled-coil (CC) highly conserved region serves as oligomerization domain for a wide variety of proteins, including structural proteins and transcription factors. The CCs typically comprise of two or more alpha helices that wrap around each other with a superficial twist. Both TIR and CC domains mainly occur in the N-terminus of NBS-LRR genes. They play an important role in R-Avr-specific recognition to trigger the downstream defense signaling response in resistant cultivars. However, TIR type occurs only in dicots whereas it is absent in monocots and has not been identified in the rice genome.

Table 7.3 Different classes of R-genes and their localization in the host cell

Sl. no.	Major R-gene classes	Localization	Example
1	Toxin reductase	Cytoplasm	*Hm1* in maize
2	NBS-LRR-CC	Cytoplasm	*RPM1* in *Arabidopsis*
3	NBS-LRR-TIR	Cytoplasm	*L6* in Flax
4	eLRR-TM	Transmembrane	*Cf-2* in tomato
5	eLRR-TM-KIN	Transmembrane	*Xa21* in Rice
6	eLRR-PEST-ECS	Transmembrane	*Ve1* in tomato
7	TM-CC	Transmembrane	*RPW8* in *Arabidopsis*
8	TIR-NBS-LRR-NLS-WRKY	Cytoplasm	*RRS1-R* in *Arabidopsis*

*Hm*1 in maize was the first disease resistance gene to be cloned and characterized (Johal and Briggs 1992), which represents a category of resistance genes that encode the detoxifying enzyme. The second major class of genes, R-genes, encodes for cytoplasmic proteins with an NBS, a C-terminal LRR, and CC domain at the N-terminus. Arabidopsis RPM1 and RPS2 and tomato I2 resistance gene belong to this class. The tobacco N-gene, flax L6 gene, and RPP5 gene are placed under the third major class of resistance genes which comprises cytoplasmic protein possessing NBS and LRR motifs and an N-terminal domain homologous to the mammalian TIR domain. The gene family devoid of NBS motif consists of extracytoplasmic leucine-rich repeats (eLRR), attached to a transmembrane (TM) domain, which is categorized under the fourth major class of resistance. eLRR plays an important role for some defense proteins like polygalacturonase-inhibiting protein (PGIPs), but they are not directly involved in pathogen recognition and activation of defense genes. Some examples of this class are *Cladosporium fulvum* resistance genes (Cf-2, Cf-4, Cf-9) of tomato and Arabidopsis FLS2. The fifth major class consists of eLRR, a transmembrane protein (TM), and cytoplasmic kinase domain (KIN). Xa21 gene in rice which provides resistance to *Xanthomonas oryzae* pv. *oryzae* represents this class. However, the eLRR in this gene makes up the most probable domain that could take part in protein-protein interactions that might affect pathogen recognition. The sixth class is comprised of those genes which have putative extracellular LRR along with a PEST (Pro-Glu-Ser-Thr) domain for protein degradation and short protein motifs for receptor-mediated endocytosis (e.g., tomato Ve1 and Ve2 genes). Arabidopsis RPW8 protein represents the seventh major class of resistance genes containing the transmembrane (TM) protein domain, fused to the coiled-coil (CC) domain. The eighth class contains Arabidopsis RRS1 R-gene which has putative nuclear localization signal (NLS) and a WRKY domain, besides TIR-NBS-LRR domains (Fukuoka et al. 2014). The WRKY domain is comprised of 60 amino acid regions defined by consensus amino acid residues WRKYGQK at the N-terminal end with a novel zinc fingerlike motif. With the progress in genome sequencing data of different plant species, list of resistance genes of this family is likely to increase.

7.4 Structure and Functions of R-Genes Conferring Rice Blast Disease

R-gene-coded proteins function in the surveillance of pathogen-coded effector proteins during infection of blast fungus in rice. Different R-gene-coded proteins confirm resistance via unique or similar mode of recognition of effector proteins.

7.4.1 R Proteins Belonging to NBS-LRR

Proteins containing NBS and LRR domains are mainly located in the cytoplasm; they are primarily responsible for recognizing effector proteins produced by the pathogen and help in downstream signaling of defense responses. The Pi37 belongs to the NBS–LRR class of R-genes and protein of Pi37 shares sequence homology more closely with the products derived from the rp1 complex proteins rather than any other R-gene sequences, yet characterized. The Pi37 gene was isolated from the St. No. 1 cultivar of rice. A map-based cloning strategy was used for the isolation of Pib gene. The deduced amino acid sequence of Pib gene containing an NBS-LRR site represents a member of the NBS-LRR class of plant disease resistance genes. The altered environmental conditions such as altered temperatures and darkness were used to induce Pib gene expression study (Lin et al. 2007a).

7.4.2 R Proteins Belonging to CC-NBS-LRR

This class of proteins along with NBS-LRR domains also contain an extra CC repeat. Ma and coworkers in 2015 have identified "Pi64 gene" as a source of rice blast resistance that confirmed resistance against both leaf and neck blast from a broad-spectrum-resistant *japonica* landrace Yangmaogu (YMG). This gene is located on chromosome 1 and encodes a CC–NBS–LRR protein. As per the expression studies, Pi64 is constitutively expressed at all the development stages, and in all tissues examined. Pish, which confers resistance against races of *M. oryzae* containing Avr-Pish, also encodes CC–NBS–LRR.

7.4.3 R Proteins Coding Other Proteins

Among the 36 characterized R-genes, 3 R-genes, *viz.* pi21, Pid-2, and Pitr, code for proline-rich heavy metal-binding protein, lectin receptor, and putative E3 ligase proteins, respectively. Product of Pitr can offer race-specific resistance to *M. oryzae* isolates carrying AVR-pita.

7.5 Characterization of Resistance Genes in Durable Blast-Resistant Rice

The blast resistance R-genes are known as Pi genes and except for three (Pi-d2, pi21, and Ptr), most of the Pi genes have sequences including both nucleotide-binding site (NBS) and leucine-rich repeat (LRR) domains (Li et al. 2019), which constitute the most prevalent class of plant resistance gene (Liu et al. 2007a). On the other hand, Pid2, pi21, and Ptr encode non-NBS-LRR protein. Pid2 is a cloned resistance gene that encodes a receptor-like kinase protein based on predicted extracellular domain of B-lectin (a bulb-type mannose-specific binding lectin) and an intracellular kinase domain (Chen et al. 2006). Pi21 gene confers non-race-specific, durable resistance, and encodes a proline-rich protein consisting of a metal-binding domain and a loss-of-function allele (pi21). However, linking of pi21 to a gene (LOC_Os04 g32890) makes it unfavorable for application because of its secondary effect on grain quality (Fukuoka et al. 2009). Out of these characterized resistance genes, seven blast R-genes along with their corresponding avirulence (Avr) genes, namely Avr-Pita and Pita (Jia et al. 2000), Avr-Pik and Pik (Kanzaki et al. 2012), Avr-Piz-t and Piz-t (Li et al. 2009), Avr-Pia and Pia (Yoshida et al. 2009), Avr1-CO39 and Pi-CO39 (Ribot et al. 2013), Avr-Pi54 and Pi54 (Vasudevan et al. 2014), and Avr-Pi9 and Pi9 (Wu et al. 2015) genes, have been comprehensively studied by researchers. Jia et al. (2000) revealed that *M. oryzae* effector, Avr-Pita, binds directly to the LRR domain of the protein Pita in both yeast two-hybrid system and *in vitro* binding assay. This presents a particular example of direct recognition of pathogen effector in rice. Pikm1-TS and Pikm2-TS at the Pikm locus of chromosome 11 (Ashikawa et al. 2008); Pikp and Pik, the multiple alleles of the Pikm locus (Yuan et al. 2011; Zhai et al. 2011); Pia locus of chromosome 11; and Pi5 locus of chromosome 9 (Lee et al. 2009b; Okuyama et al. 2011) are examples of the special feature of *M. oryzae* interaction in rice that needs two adjacent NBS-LRR genes to confer resistance. The three blast R-genes, *viz.* Pi9, Pi2, and Piz-t, are revealed to be in the same genomic region on chromosome 6 and embedded in tandemly repeated gene cluster of NBS-LRR genes (Liu et al. 2002; Qu et al. 2006; Zhou et al. 2006). *Oryza minuta*, a wild rice species, is the originator of Pi9 while Pi2 and Piz-t are allelic in nature that are from diverse local cultivars and their proteins only have changes in eight amino acids that differentiate between them (Zhou et al. 2006).

7.6 Defense-Regulator Genes as Novel Protagonists for Broad-Spectrum Blast Resistance

Significant progress has been made in tackling and mitigating the losses caused by blast disease by identifying many genes or quantitative trait loci (QTLs). Approximately 145 resistance (R) genes/alleles and 500 QTLs, related to blast resistance, have been identified. Additionally, 77 defense-related genes have also been identified and systematically studied (Li et al. 2019). They involve all defense regulators and also affect defense expression before an attack by the pathogen

(Delteil et al. 2010). Defense-regulated genes often confer partial but durable resistance against a broad range of pathogenic races. Defense-regulated genes or factors mostly belong to a transcription factor, kinases, and microRNA.

Transcription factors (TFs) are the regulatory proteins that can alter the expression of targeted genes. Being the main switches for the gene regulation mechanism, TFs act as unique candidates for targeting functional hub, dynamic networks, and nodes of different defense signaling pathways in plants. For characterizing the imperative role of TFs in disease resistance, overexpression of transgenic or downregulation is extensively adopted. However, their practical utilization leftovers were restricted in breeding programs. bsr-d1, bsr-k1, spl11, spl33, and OsBBI1 transcription factors have been reported for broad-spectrum resistance. Zhou et al. (2018) reported that rice bsr-k1 mutant exhibits durable blast resistance with no significant penalty on important agronomic traits. Spl-11 and spl-33 are identified as negative regulators for programmed cell death and defense response in rice (Zeng et al. 2004; Wang et al. 2017). Practical utilization of this transcription factor is still very much limited. Exogenous relevance of synthetic promoter and synthetic transcription factor has broadened the scope to modulate gene expression because of durable resistance over space and time.

Among the various groups of defense-regulator gene, lesion mimic mutants are one of the important members conferring resistance via mechanism lying on hypersensitive reaction, cell death, and resistance response. Lesion mimic mutant (LMM) showed disease like lesion/symptom even without pathogen invasion. spl11 (Zeng et al. 2004), spl18 (Mori et al. 2007), and lrd6-6 (Zhu et al. 2016a, b) confer blast resistance to numerous pathogenic races. However, most of the lesion mimic mutants confer blast resistance at the cost of yield. Therefore, it is a need of the hour to identify some unique mutants having durable resistance along with minimum or no yield penalty.

Worldwide, over 120,000 rice germplasms, including wild rice, represent an abundant reservoir of genetic stock that can provide a diverse range of novel R or DR genes for blast resistance. However, identification of broad-spectrum R or DR genes along with excellent agronomic traits is the most challenging task in the existing genetic stock. In crop breeding, high yield along with broad-spectrum resistance is an important goal that is followed to develop high-yielding disease-resistant rice varieties. In the present scenario, identification of new resources of defense regulators with broad-spectrum resistance against a range of pathotypes is possible by artificial mutation, site-directed mutagenesis, or novel genome editing technology.

7.7 Defense Signals Mediated by R and Defense-Regulated Genes Interweave into a Network Against *M. oryzae* Infection

R-genes code for proteins which recognize specific pathogen effectors in a specific gene-for-gene fashion. Many resistance (R) genes and quantitative trait loci (QTLs) associated with blast resistance have been identified, while only 36 blast R-genes have been successfully cloned and characterized (Ashkani et al. 2015; Yadav et al. 2019). Most of these gens encode nucleotide-binding site-leucine-rich repeat (NBS-LRR) proteins. R-genes which encode NBS-LRR proteins recognize effectors which result in the activation of effector-triggered immunity (ETI) (Jones and Dangl 2006). Small portions of these genes such as Pb1, Pi63, Pi5-1, and pi21 are induced by blast infection; however, most of the others are constitutively expressed in resistant genotypes (Lee et al. 2009b; Fukuoka et al. 2009; Xu et al. 2014; Hayashi et al. 2010a; Jia et al. 2000). Some R-genes/DR genes confer broad-spectrum resistance against blast disease. Among the seven R-genes conferring broad-spectrum resistance, wild-type Pi9, OsBBI1, Ptr, Pigm, and Pi50 positively regulate resistance while others are negative regulators. Recently, possible upstream relations between PTI and ETI defense signaling in response to blast infection by integrating the CEBiP-mediated, SPL11-mediated, and OsRac1-mediated pathways have been observed. Resistance response to blast infection can be triggered at different cellular locations. AvrPiz-t operates in the cytoplasm and nucleus while CEBiP-Spl11-OsRac1 complex transfers signals primarily in the cytoplasm and at the plasma membrane. Pid2 gene product also triggers resistance against blast in the plasma membrane (Chen et al. 2006; Li et al. 2016). Wild-type LRD6-6 inhibits activation of PR genes and accumulation of antimicrobial metabolites. In contrast, lrd6-6 causes a lesion mimic phenotype which encodes a multivesicular body (MVB)-localized AAA ATPase, and regulates the MVB-mediated vesicular trafficking (Zhu et al. 2016a, b). In host with loss of function of the Bsr-k1 gene, overexpression of OsPAL1 (accumulation of OsPAL1-7 mRNAs in the cytoplasm) promotes resistance to *M. oryzae* due to higher lignin accumulation (Zhou et al. 2018). The bsr-d1 allele confers durable and non-race-specific resistance which regulates peroxidase gene expression to achieve broad-spectrum blast resistance (Li et al. 2017). IPA1 promotes both grain yield and resistance. Non-phosphorylated IPA1 protein promotes yield while phosphorylation of IPA1 at amino acid Ser163 within its DNA-binding domain enhances blast resistance (Wang et al. 2018; Li et al. 2019). PigmR which encodes NB–LRR proteins confers broad-spectrum blast resistance accompanied by cost on the yield of rice (Deng et al. 2017). Panicle blast 1 (Pb1) is also a race-non-specific durable resistance gene for panicle blast that encodes an NB–LRR protein (Hayashi et al. 2010a). Based on subcellular locations of R and DR proteins, every cell part may involve in defense against blast, but at cellular organelle level, only a few proteins are localized. However, there is no experimental evidence for connections among various organelles (Li et al. 2019).

7.8 Conclusions and Prospect of Blast Resistance Breeding

Rice blast disease has caused devastating effects on global food security. Disease management is essentially required to sustain global food production. Keeping in view the drawbacks of chemical management, development of resistant varieties is one of the most sustainable, economic, environment-friendly approaches against the rice blast disease. Various blast-resistant R-genes have been identified and some of them are being cloned and characterized. Marker-assisted selection provides an opportunity for the incorporation of R-genes into the susceptible cultivars. With the inception of new molecular tools, it is now possible to identify and characterize the rice blast resistance R-genes and QTLs with major and minor effects. The availability of gene-based and gene-linked molecular markers helps in the introgression of resistance genes/QTLs without the influence of environmental factors. Identification of novel alleles governing the blast resistance is one of the vital exercises in the rice breeding program. Novel alleles are beneficial for the development of blast-resistant cultivars with broader adaptability. The major hindrance in the development of resistant varieties is the durability of R-genes because rice varieties with a single R-gene for specific race of pathogen become susceptible over time due to occurrence of new virulent races. The gene pyramiding or gene stacking with different R-genes with overlaying resistance spectra provides durable resistance. New molecular techniques such as global gene expression analysis, microarray, and functional genomics could assist the conventional breeding for disease resistance. Functional genomics through candidate gene identification for resistance genes has great potential in developing durable, resistant cultivars.

References

Ahn SN, Kim YK, Hong HC, Han SS, Kwon SJ, Choi HC, Moon HP, McCouch SR. Molecular mapping of a gene for resistance to a Korean isolate of rice blast. Rice Genet Newsl. 1996;13:74.

Ahn SN, Kim YK, Hong HC, Han SS, Choi HC, McCouch SR, Moon HP. Mapping of genes conferring resistance to Korean isolates of rice blast fungus using DNA markers. Korean J Breed. 1997;29(4):416–23.

Ahn SN, Kim YK, Hong HC, Han SS, Kwon SJ, Choi HC, Moon HP, McCouch SR. Molecular mapping of a new gene for resistance to rice blast (*Pyricularia grisea* Sacc.). Euphytica. 2000;116:17–22.

Amante-Bordeos A, Sitch LA, Nelson R, Damacio RD, Oliva NP, Aswidinnoor H, Leung H. Transfer of bacterial blight and blast resistance from the tetraploid wild rice *Oryza minuta* to cultivated rice. *Oryza sativa*. Theor Appl Genet. 1992;84:345–54.

Ashikawa I, Hayashi N, Yamane H, Kanamori H, Wu J, Matsumoto T, Ono K, Yano M. Two adjacent nucleotide-binding site-leucine-rich repeat class genes are required to confer *Pikm*-specific rice blast resistance. Genetics. 2008;180(4):2267–76.

Ashkani S, Yusop MR, Shabanimofrad M, Azadi A, Ghasemzadeh A, Azizi P, Latif MA. Allele mining strategies: principles and utilisation for blast resistance genes in rice (Oryza sativa L.). Curr Issues Mol Biol. 2015;17:57–74.

Ashkani S, Rafii MY, Shabanimofrad M, Ghasemzadeh A, Ravanfar SA, Latif MA. Molecular progress on the mapping and cloning of functional genes for blast disease in rice (*Oryza sativa* L.): current status and future considerations. Crit Rev Biotechnol. 2016;36:353–67.

Ballini E, Morel JB, Droc G, Price A, Courtois B, Notteghem JL, Tharreau D. A genome-wide meta-analysis of rice blast resistance genes and quantitative trait loci provides new insights into partial and complete resistance. Mol Plant-Microbe Interact. 2008;21:859–68.

Bao-hua W, Ebbole DJ, Zong-hua W. The arms race between *Magnaporthe oryzae* and rice: diversity and interaction of *Avr* and *R* genes. J Integr Agric. 2018;16(0):60345–7.

Barman SR, Gowda M, Venu RC, Chattoo BB. Identification of a major blast resistance gene in the rice cultivar Tetep. Plant Breed. 2004;123:300–2.

Berruyer R, Adreit H, Milazzo J, Gaillard S, Berger A, Dioh W, Lebrun MH, Tharreau D. Identification and fine mapping of *Pi33* the rice resistance gene corresponding to the *Magnaporthe grisea* avirulence gene *ACE1*. Theor Appl Genet. 2003;107:1139–47.

Bevitori R, Ghini R. Rice blast disease in climate change times. J Rice Res. 2014;3:e111.

Bryan GT, Wu KS, Farrall L, Jia Y, Hershey HP, McAdams SA, Faulk KN, Donaldson GK, Tarchini R, Valent B. A single amino acid difference distinguishes resistant and susceptible alleles of the rice blast resistance gene Pi-ta. Plant Cell. 2000;12(11):2033–46.

Causse MA, Fulton TM, Cho YG, Ahn SN, Chunwongse J, Wu K, Xiao J, Yu ZH, Ronald PC, Harrington SE, Second G, McCouch SR, Tanksley SD. Saturated molecular map of the rice genome based on an inter-specific backcross population. Genetics. 1994;138:1251–127.

Cesari S, Thilliez G, Ribot C, Chalvon V, Michel C, Jauneau A, Rivas S, Alaux L, Kanzaki H, Okuyama Y, Jean BM, Fournier E, Tharreau D, Terauchi R, Kroja T. The rice resistance protein pair RGA4/RGA5 recognizes the *Magnaporthe oryzae* effectors *AVR-Pia* and *AVR1-CO39* by direct binding. Plant Cell. 2013;25:1463–81.

Chauhan R, Farman M, Zhang HB, Leong S. Genetic and physical mapping of a rice blast resistance locus *Pi-CO39(t)* that corresponds to the avirulence gene *AVR1-CO39* of *Magnaporthe grisea*. Mol Genet Genomics. 2002;267:603–12.

Chen X, Temnykh S, Xu Y, Gho YG, McCouch SR. Development of a microsatellite framework map providing genome wide coverage in rice (*Oryza sativa* L.). Theor Appl Genet. 1997;95:553–67.

Chen DH, Vina MD, Inukai T, Mackill DJ, Ronald PC, Nelson RJ. Molecular mapping of the blast resistance gene *Pi44(t)* in a line derived from a durably resistant rice cultivar. Theor Appl Genet. 1999;98:1046–53.

Chen XW, Li SG, Xu JC, Zhai WX, Ling ZZ, Ma BT, Wang YP, Wang WM, Cao G, Ma YQ, Shang JJ, Zhao XF, Zhou KD, Zhu LH. Identification of two blast resistance genes in a rice variety Digu. J Phytopathol. 2004;152:77–85.

Chen S, Wang L, Que ZQ, Pan R, Pan Q. Genetic and physical mapping of *Pi37(t)*, a new gene conferring resistance to rice blast in the famous cultivar St. No. 1. Theor Appl Genet. 2005;111:1563–70.

Chen XW, Shang JJ, Chen DX, Lei CL, Zou Y, Zhai WX, Liu GZ, Xu JC, Ling ZZ, Cao G, Ma BT, Wang YP, Zhao XF, Li SG, Zhu LH. A B-lectin receptor kinase gene conferring rice blast resistance. Plant J. 2006;46:794–804.

Chen J, Shi YF, Liu W, Chai R, Fu Y, Zhuang J, Wu J. A *Pid3* allele from rice cultivar Gumei2 confers resistance to *Magnaporthe oryzae*. J Genet Genomics. 2011;38:209–16.

Chen J, Peng P, Tian J, He Y, Zhang L, Liu Z, Yin D, Zhang Z. *Pike*, a rice blast resistance allele consisting of two adjacent NBS-LRR genes, was identified as a novel allele at the *Pik* locus. Mol Breed. 2015;35:117.

Coca M, Bortolotti C, Rufat M, Pen G, Eritja R, Tharreau D, del Pozo AM, Messeguer J, Segundo BS. Transgenic rice plants expressing the antifungal AFP protein from *Aspergillus giganteus* show enhanced resistance to the rice blast fungus *Magnaporthe grisea*. Plant Mol Biol. 2004;54(2):245–59.

Das A, Soubam D, Singh PK, Thakur S, Singh NK, Sharma TR. A novel blast resistance gene, *Pi54rh* cloned from wild species of rice, *Oryza rhizomatis* confers broad spectrum resistance to *Magnaporthe oryzae*. Funct Integr Genom. 2012;12:215–28.

Dean RA, Talbot NJ, Ebbole DJ, Farman ML, Mitchell TK, Orbach MJ, Thon M, Kulkarni R, Xu JR, Pan H, Read ND, Lee YH, Carbone I, Brown D, Oh YY, Donofrio N, Jeong JS, Soanes DM, Djonovic S, Kolomiets E, Rehmeyer C, Li W, Harding M, Kim S, Lebrun MH, Bohnert H, Coughlan S, Butler J, Calvo S, Ma LJ, Nicol R, Purcell S, Nusbaum C, Galagan EJ, Birren BW. The genome sequence of the rice blast fungus *Magnaporthe grisea*. Nature. 2005;434:980–6.

Dean R, Van Kan JA, Pretorius ZA, Hammond-Kosack KE, Di Pietro A, Spanu PD, Rudd JJ, Dickman M, Kahmann R, Ellis J, Foster GD. The top 10 fungal pathogens in molecular plant pathology. Mol Plant Pathol. 2012;13:414–30.

Delteil A, Zhang J, Lessard P, Morel JB. Potential candidate genes for improving rice disease resistance. Rice. 2010;3:56–71.

Deng Y, Zhu X, Shen Y, He Z. Genetic characterization and fine mapping of the blast resistance locus Pigm (t) tightly linked to Pi2 and Pi9 in a broad-spectrum resistant Chinese variety. Theor Appl Genet. 2006;113(4):705–13.

Deng Y, Zhai K, Xie Z, Yang D, Zhu X, Liu J, Wang X, Qin P, Yang Y, Zhang G, Li Q, Zhang J, Wu S, Milazzo J, Mao B, Wang E, Xie H, Tharreau D, He Z. Epigenetic regulation of antagonistic receptors confers rice blast resistance with yield balance. Science. 2017;355:962–5.

Devanna NB, Vijayan J, Sharma TR. The blast resistance gene *Pi54of* cloned from *Oryza officinalis* interacts with *Avr-Pi54* through its novel non-LRR domains. PLoS One. 2014;9:e104840.

Dong L, Liu S, Xu P, Deng W, Li X, Tharreau D, Li J, Zhou J, Wang Q, Tao D, Yang Q. Fine mapping of *Pi57(t)* conferring broad spectrum resistance against *Magnaporthe oryzae* in introgression line IL-E1454 derived from *Oryza longistaminata*. PLoS One. 2017;12(10):e0186201.

Dwinita WU, Sugiono M, Hajrial A, Asep S, Ida H. Blast resistance genes in wild rice *Oryza rufipogon* and rice cultivar IR64. Indones J Agric Sci. 2008;1:71–6.

Ebitani T, Hayashi N, Omoteno M, Ozaki H, Yano M, Morikawa M, Fukuta Y. Characterization of *Pil3*, a blast resistance gene that maps to chromosome 6 in *indica* rice (*Oryza sativa* L. variety, Kasalath). Breed Sci. 2011;61:251–9.

Ellur RK, Khanna A, Yadav A, Pathania S, Rajashekara H, Singh VK, Krishnan SG, Bhowmick PK, Nagarajan M, Vinod KK, Prakash G. Improvement of Basmati rice varieties for resistance to blast and bacterial blight diseases using marker assisted backcross breeding. Plant Sci. 2016;242:330–41.

Endo T, Yamaguchi M, Kaji R, Nakagomi K, Yokogami M, Nakamura T, Ishikawa G, Yonimaru J, Nishio T. Close linkage of a blast resistance gene, *Pias(t),* with a bacterial leaf blight resistance gene, *Xa1-as(t)*, in a rice cultivar 'Asominori'. Breed Sci. 2012;62:334.

Fjellstrom R, Conaway-Bormans CA, McClung AM, Marchetti MA, Shank AR, Park WD. Development of DNA markers suitable for marker assisted selection of three genes conferring resistance to multiple pathotypes. Crop Sci. 2004;44:1790–8.

Fuentes LJ, Correa-Victoria FJ, Escobar F, Prado G, Aricapa G, Duque MC, Tohme J. Identification of microsatellite markers linked to the blast resistance gene *Pi-1(t)* in rice. Euphytica. 2008;160:295–304.

Fujii K, Hayano-Saito Y, Sugiura N, Hayashi N, Saka N, Tooyama T, Izawa T, Shumiya A. Gene analysis of panicle blast resistance in rice cultivars with rice stripe resistance. Breed Res. 1999;1:203–10.

Fukuoka S, Okuno K. QTL analysis and mapping of *pi21* a recessive gene for field resistance to rice blast in Japanese upland rice. Theor Appl Genet. 2001;103:85–190.

Fukuoka S, Saka N, Koga H, Ono K, Shimizu T, Ebana K, Hayashi N, Takahashi A, Hirochika H, Okuno K, Yano M. Loss of function of a proline-containing protein confers durable disease resistance in rice. Science. 2009;325(5943):998–1001.

Fukuoka S, Yamamoto SI, Mizobuchi R, Yamanouchi U, Ono K, Kitazawa N, Yasuda N, Fujita Y, Nguyen TT, Koizumi S, Sugimoto K, Matsumoto T, Yano M. Multiple functional polymorphisms in a single disease resistance gene in rice enhance durable resistance to blast. Sci Rep. 2014;4:1–7.

Fukuta Y. The *Pish* blast resistance gene of rice (*Oryza sativa* L.) is located on the long arm of chromosome 1. Japan International Research Center for Agricultural Sciences (JIRCAS). Ann Rep. 2004;46–7.

Fukuta Y, Yanoria MJT, Mercado-Escueta D, Ebron LA, Fujita Y, Araki E, Khush GS. Quantitative trait loci (QTL) reactions to rice blast isolates from Japan and the Philippines. In: Kawasaki S, editor. Rice blast: interaction with rice and control. Dordrecht: Springer; 2004. p. 113–21.

Goto I. Genetic studies on the resistance of rice plant to the blast fungus I. Inheritance of resistance in crosses Sensho x H-79 and Imochishirazu x H-79. Ann Phytopathol Soc Jpn. 1970;36:304–12.

Goto I. Genetic studies on resistance of rice plant to blast fungus II. Difference in resistance to the blast disease between Fukunishiki and its parental cultivar, Zenith. Ann Phytopathol Soc Jpn. 1976;42:253–60.

Goto I. Genetic studies on resistance of rice plant to blast fungus (VII). Blast resistance genes of Kuroka. Ann Phytopathol Soc Jpn. 1988;54:460–5.

Gowda M, Roy-Barman S, Chattoo BB. Molecular mapping of a novel blast resistance gene *Pi38* in rice using SSLP and AFLP markers. Plant Breed. 2006;125:596–9.

Gururani MA, Venkatesh J, Upadhyaya CP, Nookaraju A, Pandey SK, Park SW. Plant disease resistance genes: current status and future directions. Physiol Mol Plant Pathol. 2012;78:51–65.

Hayasaka H, Takamatsu M, Kuboki Y, Yano M, Matsunaga K, Sasaki T. Mapping genes conferring rice blast resistance in rice variety Kasalath using RFLP markers. II. Linkage analysis of the resistance gene on chromosome 6. Breed Sci. 1995;45:168.

Hayasaka H, Miyao A, Yano M, Matsunaga K, Sasaki T. RFLP mapping of a rice blast resistance gene *Pik*. Breed Sci. 1996;46:68.

Hayashi K, Yoshida H. Refunctionalization of the ancient rice blast disease resistance gene Pit by the recruitment of a retrotransposon as a promoter. Plant J. 2009;57(3):413–25.

Hayashi K, Yoshida H, Ashikawa I. Development of PCR-based allele specific and InDel marker sets for nine rice blast resistance genes. Theor Appl Genet. 2006;113(2):251–60.

Hayashi N, Ando I, Imbe T. Identification of a new resistance gene to a Chinese blast fungus isolate in the Japanese rice cultivar Aichi Asahi. Phytopathology. 1998;88:822–7.

Hayashi N, Inoue H, Kato T, Funao T, Shirota M, Shimizu T, Kanamori H, Yamane H, Hayano-Saito Y, Matsumoto T, Yano M, Takatsuji H. Durable panicle blast resistance gene *Pb1* encodes an atypical CC-NBS-LRR protein and was generated by acquiring a promoter through local genome duplication. Plant J. 2010a;64:498–510.

Hayashi K, Yasuda N, Fujita Y, Koizumi S, Yoshida H. Identification of the blast resistance gene Pit in rice cultivars using functional markers. Theor Appl Genet. 2010b;121(7):1357–67.

He X, Liu X, Wang L, Wang L, Lin F, Cheng Y, Chen Z, Liao Y, Pan Q. Identification of the novel recessive gene pi55 (t) conferring resistance to *Magnaporthe oryzae*. Sci China Life Sci. 2012;55:141–9.

Heath MC. Hypersensitive response-related death. In: Lam E, Fukuda H, Greenberg J, editors. Programmed cell death in higher plants. Dordrecht: Springer; 2000. p. 77–90.

Hittalmani S, Parco A, Mew TV, Zeigler RS, Huang N. Fine mapping and DNA marker-assisted pyramiding of the three major genes for blast resistance in rice. Theor Appl Genet. 2000;100(7):1121–8.

Hua L, Wu J, Chen C, Wu W, He X, Lin F, Wang L, Ashikawa I, Matsumoto T, Wang L, Pan Q. The isolation of *Pi1*, an allele at the *Pik* locus which confers broad spectrum resistance to rice blast. Theor Appl Genet. 2012;125(5):1047–55.

Huang H, Huang L, Feng G, Wang S, Wang Y, Liu J, Jiang N, Yan W, Xu L, Sun P, Li Z, Pan S, Liu X, Xiao Y, Liu E, Dai L, Wang GL. Molecular mapping of the new blast resistance genes Pi47 and Pi48 in the durably resistant local rice cultivar Xiangzi 3150. Phytopathology. 2011;101(5):620–6.

Imbe T, Matsumoto S. Inheritance of resistance of rice varieties to the blast fungus strains virulent to the variety “Reiho”. Jpn J Breed. 1985;35:332–9.

Imbe T, Oba S, Yanoria MJT, Tsunematsu H. A new gene for blast resistance in rice cultivar, IR24. Rice Genet Newsl. 1997;14:60–2.

Ise K. Linkage analysis of some blast resistance gene in rice, *Oryza sativa* L. Jpn J Breed. 1991;42(2):388–9.

Jeon JS, Chen D, Yi GH, Wang GL, Ronald PC. Genetic and physical mapping of *Pi5(t)* a locus associated with broad-spectrum resistance to rice blast. Mol Genet Genomics. 2003;269:280–9.

Jeung JU, Kim BR, Cho YC, Han SS, Moon HP, Lee YT, Jena KK. A novel gene *Pi40(t)* linked to the DNA markers derived from NBS-LRR motifs confers broad spectrum blast resistance in rice. Theor Appl Genet. 2007;115:1163–77.

Jia Y, McAdams SA, Bryan GT, Hershey HP, Valent B. Direct interaction of resistance gene and avirulence gene products confers rice blast resistance. EMBO J. 2000;19:4004–14.

Jiang J, Wang S. Identification of a 118 kb DNA fragment containing the locus of blast resistance gene *Pi-2(t)* in rice. Mol Genet Genomics. 2002;268:249–52.

Jiang N, Li Z, Wu J, Wang Y, Wu L, Wang S, Wang D, Wen T, Liang Y, Sun P, Liu J, Dai L, Wang Z, Wang C, Luo M, Liu X, Wang GL. Molecular mapping of the Pi2/9 allelic gene Pi2-2 conferring broad-spectrum resistance to *Magnaporthe oryzae* in the rice cultivar Jefferson. Rice. 2012;5(1):29.

Johal GS, Briggs SP. Reductase activity encoded by the HM1 disease resistance gene in maize. Science. 1992;258(5084):985–7.

Jones JDG, Dangl JL. The plant immune system. Nature. 2006;444:323–9.

Jones DA, Jones JD. The role of leucine-rich repeat proteins in plant defences. In: Advances in botanical research. Vol. 24. Academic; 1997. p. 89–167.

Joshi S, Dhatwalia S, Kaachra A, Sharma KD, Rathour R. Genetic and physical mapping of a new rice blast resistance specificity Pi-67 from a broad spectrum resistant genotype Tetep. Euphytica. 2019;215:9.

Kaji R, Ogawa T. RFLP mapping of a blast resistance gene *Pikm* in rice. Breed Sci. 1996;46(1):70.

Kanzaki H, Yoshida K, Saitoh H, Fujisaki K, Hirabuchi A, Alaux L, Fournier E, Tharreau D, Terauchi R. Arms race co-evolution of *Magnaporthe oryzae* AVR-Pik and rice Pik genes driven by their physical interactions. Plant J. 2012;72:894–907.

Khanna A, Sharma V, Ellur RK, Shikari AB, Krishnan SG, Singh UD, Prakash G, Sharma TR, Rathour R, Variar M, Prashanthi SK. Development and evaluation of near-isogenic lines for major blast resistance gene(s) in Basmati rice. Theor Appl Genet. 2015;128(7):1243–59.

Kim DM, Ju HG, Yang P, Han SS, Roh JH, Ahn SN. Mapping and race specific reaction of the resistance gene *Pi45(t)* in rice. Korean J Breed Sci. 2011;43:42–9.

Kinoshita T, Kiyosawa S. Some considerations on linkage relationships between Pii and Piz in the blast resistance of rice. Rice Genet Newsl. 1997;14:57–9.

Kiyosawa S. Genetic studies on host-pathogen relationship in the rice blast disease. Tokyo Agric Forest Fish Res Counc Trop Agric Res Ser. 1967;1:137–53.

Koide Y, Telebanco-Yanoria MJ, Pena FD, Fukuta Y, Kobayashi N. Characterization of rice blast isolates by the differential system and their application for mapping a resistance gene, *Pi19 (t)*. J Phytopathol. 2011;159:85–93.

Koide Y, Telebanco-Yanoria MJ, Fukuta Y, Kobayashi N. Detection of novel blast resistance genes *Pi58(t)* and *Pi59(t)* in a Myanmar rice landrace based on a standard differential system. Mol Breed. 2013;32:241–52.

Kumar P, Pathania S, Katoch P, Sharma TR, Plaha P, Rathour R. Genetic and physical mapping of blast resistance gene *Pi-42(t)* on the short arm of rice chromosome 12. Mol Breed. 2010;25:217–28.

Lee S, Wamishe Y, Jia Y, Liu G, Jia MH. Identification of two major resistance genes against race 1E–1k *Magnaporthe oryzae* in the indica rice cultivar Zhe733. Mol Breed. 2009a;24:127–34.

Lee S, Song MY, Seo YS, Kim HK, Ko S, Cao PJ, Suh JP, Yi G, Roh JH, Lee S, An G, Hahn TR, Wang GL, Ronald P, Jeon JS. Rice *Pi5*-mediated resistance to *Magnaporthe oryzae* requires the presence of two coiled-coil-nucleotide-binding–leucine-rich repeat genes. Genetics. 2009b;181:1627–38.

Lei CL, Huang DY, Li W, Wang LJ, Liu ZL, Wang XT, Shi K, Cheng ZJ, Zhang X, Ling ZZ, Wan JM. Molecular mapping of a blast resistance gene in an *indica* rice cultivar Yanxian No. 1. Rice Genet Newsl. 2005;22:76–7.

Lei C, Hao K, Yang Y, Ma J, Wang S, Wang J, Cheng Z, Zhao S, Zhang X, Guo X, Wang C, Wan J. Identification and fine mapping of two blast resistance genes in rice cultivar 93-11. Crop J. 2013;1:2–14.

Li LY, Wang L, Jing JX, Li ZQ, Lin F, Huang LF, Pan QH. The *Pik-m* gene, conferring stable resistance to isolates of *Magnaporthe oryzae* was finely mapped in a crossover cold region on rice chromosome 11. Mol Breed. 2007;20:179–88.

Li W, Lei C, Cheng Z, Jia Y, Huang D, Wang J, Wang J, Zhang X, Su N, Guo X, Zhai H, Wan J. Identification of SSR markers for a broad-spectrum blast resistance gene *Pi20(t)* for marker assisted breeding. Mol Breed. 2008;22:141–9.

Li W, Wang B, Wu J, Lu G, Hu Y, Zhang X, Zhang Z, Zhao Q, Feng Q, Zhang H, Wang Z, Wang G, Han B, Wang Z, Zhou B. The *Magnaporthe oryzae* avirulence gene AvrPiz-t encodes a predicted secreted protein that triggers the immunity in rice mediated by the blast resistance gene Piz-t. Mol Plant-Microbe Interact. 2009;22:411–20.

Li Y, Lu YG, Shi Y, Wu L, Xu YJ, Huang F, Guo XY, Zhang Y, Fan J, Zhao JQ, Zhang HY, Xu PZ, Zhou JM, Wu XJ, Wang PR, Wang WM. Multiple rice microRNAs are involved in immunity against the blast fungus *Magnaporthe oryzae*. Plant Physiol. 2014;164(2):1077–92.

Li W, Liu Y, Wang J, He M, Zhou X, Yang C, Yuan C, Wang J, Chern M, Yin J, Yin J, Chen W, Ma B, Wang Y, Qin P, Li S, Ronald P, Chen X. The durably resistant rice cultivar Digu activates defence gene expression before the full maturation of Magnaporthe oryzae appressorium. Mol Plant Pathol. 2016;17:354–68.

Li W, Zhu Z, Chern M, Yin J, Yang C, Ran L, Cheng M, He M, Wang K, Wang J, Zhou X, Zhu X, Chen Z, Wang J, Zhao W, Ma B, Qin P, Chen W, Wang Y, Liu J, Wang W, Wu X, Li P, Wang J, Zhu L, Li S, Chen X. A natural allele of a transcription factor in rice confers broad-spectrum blast resistance. Cell. 2017;170:114–26.

Li W, Chern M, Yin J, Wang J, Chen X. Recent advances in broad-spectrum resistance to the rice blast disease. Curr Opin Plant Biol. 2019;50:114–20.

Liang Z, Wang L, Pan Q. A new recessive gene conferring resistance against rice blast. Rice. 2016;9:47.

Lin F, Chen S, Que Z, Wang L, Liu X, Pan Q. The blast resistance gene *Pi37* encodes a nucleotide binding site leucine-rich repeat protein and is a member of a resistance gene cluster on rice chromosome 1. Genetics. 2007a;177:1871–80.

Lin LY, Jing JX, Li ZQ, Lin F, Huang LF, Pan QH. A high resolution map of the rice blast resistance gene *Pi15* constructed by sequence-ready markers. Plant Breed. 2007b;126:287–90.

Liu G, Lu G, Zeng L, Wang GL. Two broad spectrum blast resistance genes, Pi9 (t) and Pi2 (t), are physically linked on rice chromosome 6. Mol Gen Genomics. 2002;267:472–80.

Liu B, Zhang S, Zhu X, Yang Q, Wu S, Mei M, Mauleon R, Leach J, Mew T, Leung H. Candidate defense genes as predictors of quantitative blast resistance in rice. Mol Plant-Microbe Interact. 2004;17:1146–52.

Liu XQ, Wang L, Chen S, Lin F, Pan QH. Genetic and physical mapping of *Pi36(t)*, a novel rice blast resistance gene located on rice chromosome 8. Mol Gen Genomics. 2005;274:394–401.

Liu J, Liu X, Dai L, Wang GL. Recent progress in elucidating the structure, function and evolution of disease resistance genes in plants. J Genet Genomics. 2007a;34:765–76.

Liu X, Lin F, Wang L, Pan Q. The in silico map-based cloning of Pi36, a rice coiled-coil–nucleotide-binding site–leucine-rich repeat gene that confers race-specific resistance to the blast fungus. Genetics. 2007b;176(4):2541–9.

Liu X, Yang Q, Lin F, Hua L, Wang C, Wang L, Pan Q. Identification and fine mapping of Pi39(t), a major gene conferring the broad-spectrum resistance to *Magnaporthe oryzae*. Mol Gen Genomics. 2007c;278:403–10.

Liu Y, Liu B, Zhu X, Yang J, Bordeos A, Wang G, Leach JE, Leung H. Fine-mapping and molecular marker development for Pi56 (t), a NBS-LRR gene conferring broad spectrum resistance to *Magnaporthe oryzae* in rice. Theor Appl Genet. 2012;126:985–98.

Lü Q, Xu X, Shang J, Jiang G, Pang Z, Zhou Z, Wang J, Liu Y, Li T, Li X, Xu J, Cheng Z, Zhao X, Li S, Zhu L. Functional analysis of *Pid3-A4*, an ortholog of rice blast resistance gene *Pid3* revealed by allele mining in common wild rice. Phytopathology. 2013;103:594–9.

Ma J, Lei C, Xu X, Hao K, Wang J, Cheng Z, Ma X, Ma J, Zhou K, Zhang X, Guo X, Wu F, Lin Q, Wang C, Zhai H, Wang H, Wan J. *Pi64*, encoding a novel CC-NBS-LRR protein, confers resistance to leaf and neck blast in rice. Mol Plant-Microbe Interact. 2015;28(5):558–68.

Mackill DJ, Bonman JM. Inheritance of blast resistance in near-isogenic lines of rice. Phytopathology. 1992;82:746–9.

Marathi B, Jena KK. Floral traits to enhance outcrossing for higher hybrid seed production in rice: present status and future prospects. Euphytica. 2015;201:1–14.

McCouch SR, Kochert G, Yu ZH, Wang ZY, Khush GS, Coffman WR, Tanksley SD. Molecular mapping of rice chromosomes. Theor Appl Genet. 1988;76(6):815–29.

McCouch SR, Nelson RJ, Tohme J, Zeigler RS. Mapping of blast resistance genes in rice. In: Zeigler RS, Leong SA, Teng PS, editors. Rice blast disease. Wallingford: C.A.B. International; 1994. p. 167–86.

Monna L, Miyao A, Inoue T, Fukuoka S, Yamazaki M, Zhong HS, Sasaki T, Minobe Y. Determination of RAPD markers in rice and their conversion into sequence tagged sites (STSs) and STS-specific primers. DNA Res. 1994;1:139–48.

Mori M, Tomita C, Sugimoto K, Hasegawa M, Hayashi N, Dubouzet JG, Ochiai H, Sekimoto H, Hirochika H, Kikuchi S. Isolation and molecular characterization of a spotted leaf 18 mutant by modified activation-tagging in rice. Plant Mol Biol. 2007;63:847–60.

Naqvi NI, Chattoo BB. Molecular genetic analysis and sequence characterized amplified region-assisted selection of blast resistance in rice. Rice Genet. 1996;3:570–6.

Naqvi NI, Bonman JM, Mackill DJ, Nelson RJ, Chattoo BB. Identification of RAPD markers linked to a major blast resistance gene in rice. Mol Breed. 1995;1:341–8.

Nguyen TTT, Koizumi S, La TN, Zenbayashi KS, Ashizawa T, Yasuda N, Imazaki I, Miyasaka A. *Pi35*(t), a new gene conferring partial resistance to leaf blast in the rice cultivar Hokkai 188. Theor Appl Genet. 2006;113(4):697–704.

Okuyama Y, Kanzaki H, Abe A, Yoshida K, Tamiru M, Saitoh H, Fujibe T, Matsumura H, Shenton M, Galam DC, Undan J, Ito A, Sone T, Terauchi R. A multifaceted genomics approach allows the isolation of the rice *Pia*-blast resistance gene consisting of two adjacent NBS-LRR protein genes. Plant J. 2011;66(3):467–79.

Pan Q, Wang L, Ikehashi H, Tanisaka T. Identification of a new blast resistance gene in the *indica* rice cultivar Kasalath using Japanese differential cultivars and isozyme markers. Phytopathology. 1996a;86:1071–5.

Pan QH, Tanisaka T, Ikehashi H. Studies on the genetics and breeding of blast resistance in rice VI. Gene analysis for the blast resistance of two Yunnan native cultivars GA20 and GA25. Breed Sci. 1996b;46(2):70.

Pan Q, Tanisaka T, Ikehashi H. Studies on the genetics and breeding of blast resistance in rice VII. Gene analysis for the blast resistance of Indian native cultivar, Aus 373. Breed Sci. 1997;47:35.

Pan QH, Wang L, Ikehashi H, Yamagata H, Tanisaka T. Identification of two new genes conferring resistance to rice blast in the Chinese native cultivar 'Maowangu'. Plant Breed. 1998;117:27–31.

Pan QH, Wang L, Tanisaka T. A new blast resistance gene identified in the Indian native rice cultivar Aus373 through allelism and linkage tests. Plant Pathol. 1999;48(2):288–93.

Pan QH, Hu ZD, Tanisaka T, Wang L. Fine mapping of the blast resistance gene Pi15, linked to Pii, on rice chromosome 9. Acta Bot Sin. 2003;45:871–7.

Parco A. STS markers for the blast resistance gene *Pi-1(t)*. Rice Genome. 1995;4:9.

Qu S, Liu G, Zhou B, Bellizzi M, Zeng L, Dai L, Han B, Wang GL. The broad-spectrum blast resistance gene Pi9 encodes a nucleotide-binding site–leucine-rich repeat protein and is a member of a multigene family in rice. Genetics. 2006;172(3):1901–14.

Rani R, Yadav P, Barbadikar KM, Baliyan N, Malhotra EV, Singh BK, Kumar A, Singh D. CRISPR/Cas9: a promising way to exploit genetic variation in plants. Biotechnol Lett. 2016;11(1):1–8.

Ribot C, Cesari S, Abidi I, Chalvon V, Bournaud C, Vallet J, Lebrun M, Morel J, Kroj T. The *Magnaporthe oryzae* effector AVR1CO39 is translocated into rice cells independently of a fungal-derived machinery. Plant J. 2013;74:1–12.

Rybka K, Miyamoto M, Ando I, Saito A, Kawasaki S. High resolution mapping of the *indica*-derived rice blast resistance genes II. *Pi-ta2* and *Pi-ta* and a consideration of their origin. Mol Plant-Microbe Interact. 1997;10:517–24.

Sallaud C, Lorieux M, Roumen E, Tharreau D, Berruyer R, Svestasrani P, Garsmeur O, Ghesquiere A, Notteghem JL. Identification of five new blast resistance genes in the highly blast resistant rice variety IR64 using a QTL mapping strategy. Theor Appl Genet. 2003;106:794–803.

Sasaki R. Existence of strains in rice blast fungus. Int J Plant Prot Tokyo. 1922;9:631–44.

Seck PA, Diagne A, Mohanty S, Wopereis MCS. Crops that feed the world. Food Secur. 2012;4:7–24.

Shang J, Tao Y, Chen X, Zou Y, Lei C, Wang J, Li X, Zhao X, Zhang M, Lu Z, Xu J, Cheng Z, Wan J, Zhu L. Identification of a new rice blast resistance gene, Pid3, by genome-wide comparison of paired NBS-LRR genes and their pseudogene alleles between the two sequenced rice genomes. Genetics. 2009;182(4):1303–11.

Sharma TR, Madhav MS, Singh BK, Shanker P, Jana TK, Dalal V, Pandit A, Singh A, Gaikwad K, Upreti HC, Singh NK. High-resolution mapping, cloning and molecular characterization of the Pi-k h gene of rice, which confers resistance to *Magnaporthe grisea*. Mol Gen Genomics. 2005;274(6):569–78.

Sharma TR, Rai AK, Gupta SK, Singh NK. Broad-spectrum blast resistance gene *Pi-kh* cloned from rice line Tetep designated as *Pi54*. J Plant Biochem Biotechnol. 2010;19:87–9.

Shi BH, Zhang JH, Zheng YM, Liu YQ, Cruz CV, Zheng TQ, Zhao MF. Identification of a new resistance gene Pi-Da (t) from Dacca6 against rice blast fungus (*Magnaporthe oryzae*) in Jin23B background. Mol Breed. 2012;30:1089–96.

Shinoda H, Toriyama K, Yunoki T, Ezuka A, Sakurai Y. Studies on the varietal resistance of rice to blast. VI. Linkage relationship of blast resistance genes. Bull Chugoku Nat Agric Exp Stat A. 1971;20:1–25.

Singh VK, Singh A, Singh SP, Ellur RK, Choudhary V, Sarkel S, Singh D, Gopala KS, Nagarajan M, Vinod KK, Singh UD, Prabhu KV, Sharma TR, Mohapatara T, Singh AK. Incorporation of blast resistance into "PRR78", an elite Basmati rice restorer line, through marker assisted backcross breeding. Field Crops Res. 2012;128(1):8–16.

Singh WH, Kapila RK, Sharma TR, Rathour R. Genetic and physical mapping of a new allele of *Pik* locus from *japonica* rice 'Liziangxintuanheigu'. Euphytica. 2015;205:889–901.

Spence C, Alff E, Johnson C, Ramos C, Donofrio N, Sundaresan V, Bais H. Natural rice rhizospheric microbes suppress rice blast infections. BMC Plant Biol. 2014;14:130.

Srivastava D, Md S, Kumar M, Mishra A, Pandey P, Kumar D, Yadav P, Md HS, Singh KN. Current status of conventional and molecular interventions for blast resistance in rice. Rice Sci. 2017;24(6):299–321.

Su J, Wang W, Han J, Chen S, Wang C, Zeng L, Feng A, Yang J, Zhou B, Zhu X. Functional divergence of duplicated genes results in a novel blast resistance gene *Pi50* at the *Pi2/9* locus. Theor Appl Genet. 2015;128:2213–25.

Sun P, Liu J, Wang Y, Jiang N, Wang S, Dai Y, Gao J, Li Z, Pan S, Wang D, Li W, Liu X, Xiao Y, Liu E, Wang GL, Dai L. Molecular mapping of the blast resistance gene *Pi49* in the durably resistant rice cultivar Mowanggu. Euphytica. 2013;192:45–54.

Tabien RE, Pinson SRM, Marchetti MA, Li Z, Park WD, Paterson AH, Stansel JW. Blast resistance genes from Teqing and Lemont. In: Khush GS, editor. Rice genetics III. Proceedings of Third International Rice Genetics Symposium, 16–20 Oct, Manila, The Philippines; 1996. p. 451–5.

Tabien RE, Li Z, Paterson AH, Marchetti MA, Stansel JW, Pinson SRM. Mapping of four rice blast resistance genes from Lemont and Teqing and evaluation of their combinatorial effect for field resistance. Theor Appl Genet. 2000;101:1215–25.

Tabien R, Li Z, Paterson A, Marchetti M, Stansel J, Pinson S. Mapping QTLs for field resistance to the rice blast pathogen and evaluating their individual and combined utility in improved varieties. Theor Appl Genet. 2002;105:313–24.

Takagi H, Uemura A, Yaegashi H, Tamiru M, Abe A, Mitsuoka C, Utsushi H, Natsume S, Kanzaki H, Matsumura H, Saitoh H, Yoshida K, Cano LM, Kamoun S, Terauchi R. MutMap-Gap: whole-genome resequencing of mutant F2 progeny bulk combined with de novo assembly of gap regions identifies the rice blast resistance gene *Pii*. New Phytol. 2013;200:276–83.

Takahashi A, Hayashi N, Miyao A, Hirochika H. Unique features of the rice blast resistance Pish locus revealed by large scale retrotransposon-tagging. BMC Plant Biol. 2010;10(1):175.

Takehisa H, Yasuda M, Fukuta Y, Kobayashi N, Hayashi N, Nakashita H, Abe T, Sato T. Genetic analysis of resistance genes in an Indica-type rice (*Oryza sativa* L.), Kasalath, using DNA markers. Breed Sci. 2009;59:253–60.

Talbot NJ. On the trail of a cereal killer: exploring the biology of *Magnaporthe grisea*. Annu Rev Microbiol. 2003;57(1):177–202.

Terashima T, Fukuoka S, Saka N, Kudo S. Mapping of a blast field resistance gene *Pi39(t)* of elite rice strain Chubu 111. Plant Breed. 2008;127:485–9.

Traut TW. The functions and consensus motifs of nine types of peptide segments that form different types of nucleotide-binding sites. In: EJB reviews. Berlin, Heidelberg: Springer; 1994. p. 105–15.

Vasudevan K, Vera CC, Gruissem W, Bhullar NK. Large scale germplasm screening for identification of novel rice blast resistance sources. Front Plant Sci. 2014;5:505.

Wang GL, Mackill DJ, Bonman JM, McCouch SR, Champoux MC, Nelson RJ. RFLP mapping of genes conferring complete and partial resistance to blast in a durably resistant rice cultivar. Genetics. 1994;136:1421–34.

Wang Z, Yano M, Yamanouchi U, Iwamoto M, Monna L, Hayasaka H, Katayose Y, Sasaki T. The *Pib* gene for rice blast resistance belongs to the nucleotide binding and leucine-rich repeat class of plant disease resistance genes. Plant J. 1999;19:55–64.

Wang L, Xu X, Lin F, Pan Q. Characterization of rice blast resistance genes in the Pik cluster and fine mapping of the Pik-p locus. Phytopathology. 2009;99(8):900–5.

Wang Y, Wang D, Deng X, Liu J, Sun P, Liu Y, Huang H, Jiang N, Kang H, Ning Y, Wang Z, Xiao Y, Liu X, Liu E, Dai L, Wang GL. Molecular mapping of the blast resistance genes *Pi2-1* and *Pi51(t)* in the durably resistant rice Tianjingyeshengdao. Phytopathology. 2012;102:779–86.

Wang D, Guo CJ, Huang J, Yang SH, Tian DC, Zhang XH. Allele mining of rice blast resistance genes at *AC134922* locus. Biochem Biophys Res Commun. 2014;446(4):1085–90.

Wang R, Fang W, Guan C, He W, Bao Y, Zhang H. Characterization and fine mapping of a blast resistant gene *Pi-jnw1* from the *japonica* rice landrace Jiangnanwan. PLoS One. 2016;11(12):e0169417.

Wang S, Lei C, Wang J, Ma J, Tang S, Wang C, Zhao K, Tian P, Zhang H, Qi C, Cheng Z, Zhang X, Guo X, Liu L, Wu C, Wan J. SPL33, encoding an eEF1A-like protein, negatively regulates cell death and defense responses in rice. J Exp Bot. 2017;68(5):899–913.

Wang J, Zhou L, Shi H, Chern M, Yu H, Yi H, He M, Yin J, Zhu X, Li Y, Li W, Liu J, Wang J, Chen X, Qing H, Wang Y, Liu G, Wang W, Li P, Wu X, Zhu L, Zhou JM, Ronald PC, Li S, Li J, Chen X. A single transcription factor promotes both yield and immunity in rice. Science. 2018;361:1026–8.

Wu KS, Tanksley SD. Abundance, polymorphism and genetic mapping of microsatellites in rice. Mol Gen Genet. 1993;241(1–2):225–35.

Wu KS, Martinez C, Lentini Z, Tohme J, Chumley FG, Scolnik PA, Valent B. Cloning a blast resistance gene by chromosome walking. In: Khush GS, editor. Proceedings of the Third International Rice Genetics Symposium, IRRI, Manila, Philippines; 1996. p. 669–74.

Wu JL, Fan YY, Li DB, Zheng KL, Leung H, Zhuang JY. Genetic control of rice blast resistance in the durably resistant cultivar Gumei 2 against multiple isolates. Theor Appl Genet. 2005;111:50–6.

Wu YY, Bao YM, Xie LJ, Su YY, Chu RZ, He WW, Huang J, Wang JF, Zhang HS. Fine mapping and identification of blast resistance gene *Pi-hk1* in a broad-spectrum resistant *japonica* rice landrace. Phytopathology. 2013;103(11):1162–8.

Wu J, Kou Y, Bao J, Li Y, Tang M, Zhu X, Ponaya A, Xiao G, Li J, Li C, Song MY, Cumagun CJ, Deng Q, Lu G, Jeon JS, Naqvi NI, Zhou B. Comparative genomics identifies the *Magnaporthe oryzae* avirulence effector AvrPi9 that triggers Pi9-mediated blast resistance in rice. New Phytol. 2015;206:1463–75.

Xiao W, Yang Q, Wang H, Guo T, Liu Y, Zhu X, Chen Z. Identification and fine mapping of a resistance gene to *Magnaporthe oryzae* in a space-induced rice mutant. Mol Breed. 2011;28:303–12.

Xiao W, Yang Q, Wang H, Duan J, Guo T, Liu Y, Zhu X, Chen Z. Identification and fine mapping of a major R gene to *Magnaporthe oryzae* in a broad-spectrum resistant germplasm in rice. Mol Breed. 2012;30:1715–26.

Xiao W, Yang Q, Sun D, Wang H, Guo T, Liu Y, Zhu X, Chen Z. Identification of three major R genes responsible for broad-spectrum blast resistance in an *indica* rice accession. Mol Breed. 2015;35:49.

Xiu-Ying H, Liu X, Wang L, Wang L, Lin F, Cheng Y, Pan Q. Identification of the novel recessive gene *pi55(t)* conferring resistance to *Magnaporthe oryzae*. Sci China Life Sci. 2012;55(2):141–9.

Xu X, Chen H, Fujimura T, Kawasaki S. Fine mapping of a strong QTL of field resistance against rice blast *Pikahei-1(t)* from upland rice Kahei utilizing a novel resistance evaluation system in the greenhouse. Theor Appl Genet. 2008a;117:997–1008.

Xu X, Hayashi N, Wang CT, Kato H, Fujimura T, Kawasaki S. Efficient authentic fine mapping of the rice blast resistance gene *Pik-h* in the *Pik* cluster, using new *Pik-h* differentiating isolates. Mol Breed. 2008b;22:289–99.

Xu X, Hayashi N, Wang C, Fukuoka S, Kawasaki S, Takatsuji H, Jiang C. Rice blast resistance gene Pikahei-1(t), a member of a resistance gene cluster on chromosome 4, encodes a nucleotide-binding site and leucine-rich repeat protein. Mol Breed. 2014;34:691–700.

Yadav MK, Aravindan S, Ngangkham U, Raghu S, Prabhukarthikeyan SR, Keerthana U, Marndi BC, Adak T, Munda S, Deshmukh R, Pramesh D. Blast resistance in Indian rice landraces: genetic dissection by gene specific markers. PLoS One. 2019;14(1):e0211061.

Yang Q, Lin F, Wang L, Pan Q. Identification and mapping of *Pi41* a major gene conferring resistance to rice blast in the *Oryza sativa* subspecies *indica* reference cultivar 93-11. Theor Appl Genet. 2009;118:1027–34.

Yoshida K, Saitoh H, Fujisawa S, Kanzaki H, Matsumura H, Yoshida K, Tosa Y, Chuma I, Takano Y, Win J, Kamoun S, Terauchi R. Association genetics reveals three novel avirulence genes from the rice blast fungal pathogen Magnaporthe oryzae. Plant Cell. 2009;21(5):1573–91.

Yu ZH, Mackill DJ, Bonman JM, Tanksley SD. Tagging genes for blast resistance in rice via linkage to RFLP markers. Theor Appl Genet. 1991;81(4):471–6.

Yu ZH, Mackill DJ, Bonman JM, McCouch SR, Guiderdoni E, Notteghem JL, Tanksley SD. Molecular mapping of genes for resistance to rice blast (*Pyricularia grisea* Sacc.). Theor Appl Genet. 1996;93:859–63.

Yuan B, Zhai C, Wang W, Zeng X, Xu X, Hu H, Lin F, Wang L, Pan Q. The Pik-p resistance to *Magnaporthe oryzae* in rice is mediated by a pair of closely linked CC-NBS-LRR genes. Theor Appl Genet. 2011;122(5):1017–28.

Zenbayashi K, Fukuoka S, Katagiri S, Fujisawa M, Matsumoto T, Ashizawa T, Koizumi S. Genetic and physical mapping of the partial resistance gene *Pi34* to blast in rice. Phytopathology. 2007;97:598–602.

Zeng LR, Qu S, Bordeos A, Yang C, Baraoidan M, Yan H, Xie Q, Nahm BH, Leung H, Wang GL. Spotted leaf11, a negative regulator of plant cell death and defense, encodes a U-box/armadillo repeat protein endowed with E3 ubiquitin ligase activity. Plant Cell. 2004;16:2795–808.

Zeng XS, Yang XF, Zhao ZH, Lin F, Wang L, Pan Q. Characterization and fine mapping of the rice blast resistance gene *Pia*. Sci China Life Sci. 2011;54:372–8.

Zhai C, Lin F, Dong ZQ, He XY, Yuan B, Zeng XS, Wang L, Pan QH. The isolation and characterization of *Pik*, a rice blast resistance gene which emerged after rice domestication. New Phytol. 2011;189:321–34.

Zhai C, Zhang Y, Yao N, Lin F, Liu Z, Dong Z, Wang L, Pan Q. Function and interaction of the coupled genes responsible for *Pik-h* encoded rice blast resistance. PLoS One. 2014;9:e98067.

Zhang J, Li X, Jiang G, Xu Y, He YQ. Pyramiding of *Xa7* and *Xa21* for the improvement of disease resistance to bacterial blight in hybrid rice. Plant Breed. 2006;125(6):600–5.

Zhang HJ, Li GJ, Li W, Song FM. Transgenic strategies for improving rice disease resistance. Afr J Biotechnol. 2009;8(9):1750–7.

Zhao H, Wang X, Jia Y, Minkenberg B, Wheatley M, Fan J, Jia MH, Famoso A, Edwards D, Wamishe Y, Valent B, Wang GL, Yang Y. The rice blast resistance gene *Ptr* encodes an atypical protein required for broad-spectrum disease resistance. Nat Commun. 2018;9(1):2039.

Zheng W, Wang Y, Wang L, Ma Z, Zhao J, Wang P, Zhang L, Liu Z, Lu X. Genetic mapping and molecular marker development for *Pi65(t)*, a novel broad-spectrum resistance gene to rice blast using next-generation sequencing. Theor Appl Genet. 2016;129:1035–44.

Zhou JH, Wang JL, Xu JC, Lei CL, Ling ZZ. Identification and mapping of a rice blast resistance gene *Pi-g(t)* in the cultivar Guangchangzhan. Plant Pathol. 2004;53(2):191–6.

Zhou B, Qu S, Liu G, Dolan M, Sakai H, Lu G, Bellizzi M, Wang GL. The eight amino acid differences within three leucine-rich repeats between Pi2 and Piz - t resistance proteins determine the resistance specificity to *Magnaporthe grisea*. Mol Plant-Microbe Interact. 2006;19:1216–28.

Zhou X, Liao H, Chern M, Yin J, Chen Y, Wang J, Zhu X, Chen Z, Yuan C, Zhao W, Wang J, Li W, He M, Ma B, Wang J, Qin P, Chen W, Wang Y, Liu J, Qian Y, Wang W, Wu X, Li P, Zhu L, Li S, Ronald PC, Chen X. Loss of function of a rice TPR-domain RNA-binding protein confers broad-spectrum disease resistance. Proc Natl Acad Sci U S A. 2018;115(12):3174–9.

Zhu M, Wang L, Pan Q. Identification and characterization of a new blast resistance gene located on rice chromosome 1 through linkage and differential analyses. Phytopathology. 2004;94:515–9.

Zhu YY, Fang H, Wang YY, Fan JX, Yang SS, Mew TW, Mundt CC. Panicle blast and canopy moisture in rice cultivar mixtures. Phytopathology. 2005;95(4):433–8.

Zhu X, Chen S, Yang J, Zhou S, Zeng L, Han J, Jing S, Ling W, Pan Q. The identification of *Pi50(t)*, a new member of the rice blast resistance *Pi2/Pi9* multigene family. Theor Appl Genet. 2012;124:1295–304.

Zhu D, Kang H, Li Z, Liu M, Zhu X, Wang Y, Wang D, Wang Z, Liu W, Wang G-L. A genome-wide association study of field resistance to *Magnaporthe oryzae* in rice. Rice. 2016a;9:44.

Zhu X, Yin J, Liang S, Liang R, Zhou X, Chen Z, Zhao W, Wang J, Li W, He M, Yuan C, Miyamoto K, Ma B, Wang J, Qin P, Chen W, Wang Y, Wang W, Wu X, Yamane H, Zhu L, Li S, Chen X. The multivesicular bodies (MVBs)-localized AAA ATPase LRD6-6 inhibits immunity and cell death likely through regulating MVBs-mediated vesicular trafficking in rice. PLoS Genet. 2016b;12(9):e1006311.

Zhuang JY, Lu J, Qian HR, Lin HX, Zheng KL. Tagging of blast resistance gene(s) to DNA markers and marker-assisted selection (MAS) in rice improvement. In: IAEA-TECDOC:1010, Int Atomic Energy Agency (IAEA); 1998. p. 55–61.

Zhuang JY, Wu JL, Fan YY, Rao ZM, Zheng KL. Genetic drag between a blast resistance gene and QTL conditioning yield trait detected in a recombinant inbred line population in rice. Rice Genet Newsl. 2001;18:69–70.

Zhuang JY, Ma WB, Wu JL, Chai RY, Lu J, Fan YY, Jin MZ, Leung H, Zheng KL. Mapping of leaf and neck blast resistance genes with resistance gene analog, RAPD and RFLP in rice. Euphytica. 2002;128(3):363–70.

Chapter 8
Blast Disease of Rice: Evolution and Adaptation in Context of Changing Climate

Rashmi Singh and Sudarshan Maurya

Contents

8.1 Introduction

Food security is becoming a global challenge in the era of elevating climate change scenario, in the context of rice, which has driven an increased focus on developed and improved technologies of crop protection to cope up biotic and abiotic stresses. Rice (*Oryza sativa*) is the primary and staple food consumed daily by more than 50% of the world's population. Extreme weather events in climate change, combined with increased air temperature and atmospheric CO_2 concentration, are anticipated to spread diseases of rice in fresh neighbourhood (Anderson et al. 2004). Biotic stresses, viz. fungi, bacteria, viruses and nematodes, can infect more or less in the cropping season of rice and cause significant biological yield losses. Among the various biotic stresses of rice, rice blast or rotten neck blast is considered as a major yield-influencing fungal disease of the rice-growing countries of the world.

R. Singh
Smt. Indira Gandhi Post Graduate Degree College, Lalganj, Mirzapur, Uttar Pradesh, India

S. Maurya (✉)
ICAR-Indian Institute of Vegetable Research, Varanasi, India

S. C. Nayaka et al. (eds.), *Blast Disease of Cereal Crops*, Fungal Biology,
https://doi.org/10.1007/978-3-030-60585-8_8

The disease causes significant economic yield losses annually, and each year it is estimated to wipe out a huge amount of rice which is quite enough to feed more than 60 million people of the world. The blast disease was first documented in 1637 in China, and then it spread from China to Japan in 1704, and then after that it was reported in almost all the rice-growing countries gradually, in Italy (1828), the USA (1886) and India (1913), by Veeraraghavan and Padmanabhan (1965). In India, blast disease was more or less consistently recorded in major rice-growing areas. The fungus is now known to be prevalent in more than 85 countries worldwide. Rice blast is caused by *Magnaporthe oryzae*, which was differentiated from *Magnaporthe grisea* based on multilocus gene sequence-based phylogenetic analysis (Couch and Kohn 2002). Better understanding of rice blast pathosystem which consists of two inter-related subsystems, leaf blast and neck blast pathosystem, is needed for effective management of the disease (Teng et al. 1991; Teng 1994; Savary et al. 2006, 2012). Within the subsystem, vertical and horizontal host resistance governs the host resistance against future infection. The understanding of infection is very important in non-host and host plants (rice) that will be helpful in rice blast forecasting and disease management (Padmanabhan 1965). In most of the infection in subsystem it is thought to occur with rich inoculums from rice plants in their immediate vicinity, which have been successfully infected, or from the pathogen of non-hosts. Once infection has been established with an initial amount of disease, then further disease severity increases through secondary spread.

8.2 Evolution of *Pyricularia*

Pyricularia, a genus of pathogen, has a very high evolutionary potential in the aberrant climatic condition. The evolutionary potential of a pathogen population reflects its ecology and biology, and its population genetic structure (McDonald and Linde 2002). Knowledge about the evolutionary potential of *Pygt* populations is needed to predict the durability of genetic resistance to wheat blast. An intense search for blast resistance began with the first report of the disease more than 30 years ago but breeding success has been erratic and inconsistent. The average durability of resistant wheat varieties has been only 2–3 years. Furthermore, wheat genotypes behaved differently in different regions, indicating genotype-by-environment interactions or region-specific distribution of virulence groups. Reports indicated that *Pygt* is present in all Brazilian wheat-growing areas; it is likely that both the incidence and severity of wheat blast are affected by the virulence groups that predominate in each region. In fact, the occurrence of virulence groups in *Pygt* populations was already described, but information about the virulence composition and genetic structure of contemporary populations of the wheat blast pathogen remains limited.

8.3 Adaptation

That any organism has adapted to its habitat means that it has evolved diverse molecular mechanisms that allow it to grow optimally spatio-temporally with altering physico-chemical conditions and their environment (Katsantonis et al. 2017). Each organism runs after their optimum fitness in the changing environments; there are delicate differences in fitness between individuals (due to genomic plasticity or metabolic flexibility) and phylogenetic complexity (numbers and diversity of species within a given community) which can lead to the diversification of species or the extinction of less fitted genotypes over time (Kassen 2009). Aaron et al. (2010) had given emphasis on the environmental adaptation of the microorganism on three evolutionary perspectives, i.e. (1) acclimation of the existing cellular machinery to operate optimally in a new environmental niche, (2) acquisition of entirely new capabilities through horizontal gene transfer or neo-functionalisation of gene duplications and (3) reorganisation of network dynamics to appropriately adjust existing physiological processes to match dynamic environmental changes. An environment is extremely heterogeneous at microscales, and microorganisms are challenged by fluctuating biotic and abiotic stresses and parameter mixed up in changes in pH (Hughes et al. 2007), in inter- and intraspecific competition and in nutrient and resource availability (Chesson 2000). An important illustration was made by Mitchell et al. (2009) in the study of *Escherichia coli* in the digestive tract of mammals that went through a succession of carbon sources such as lactose and maltose and a succession of stresses such as increasing temperature and decreasing oxygen levels. These changes occur in a definite time and space and vary infrequently, for instance within the time frame of a single generation. Likewise adaptations can occur via several mechanisms, such as an increase in measured gene quantity, neo-functionalisation and sub-functionalisation. In all these mechanisms, mutational changes in coding regions, changes in gene expression or a grouping of both drives adaptation. Reports in *Pseudomonas fluorescens* SBW25 have also smartly demonstrated the evolution of novel phenotypes in vitro (Beaumont et al. 2009).

8.4 Changing Climate

Global population is increasing rapidly and the availability of natural resources for crop production continues to decline day by day which is escalating the challenge of global food security. An anticipated world's population of humans will be nine billion by 2050 and this is challenged by a shrinking of major land for rice (*Oryza sativa* L.) production, which is expected to decline by 18–51% in the tropics during the next century due to global warming (Godfray et al. 2010). Climate change directly or indirectly influences all the agricultural crops including cereal production through abiotic and biotic stresses, viz. heat stress, water stress along with waterlogging, frost, disease and pest infestations (Porter et al. 2014). Challinor et al. (2014) predicted a decline of the yields of wheat, maize and rice in tropical and temperate regions. Baker (2004) reported the effect of elevated CO_2 (700 μmol mol^{-1}) under different temperature regimes at different temperatures, viz. 24, 28, 32, 36 and 40 °C, and found no increase in rice grains. On the contrary, Yang et al. (2006) showed that elevating the concentration of atmospheric CO_2 increased rice productivity. Goria et al. (2013) showed that the deleterious effect of elevated carbon dioxide concentration is likely to modify plant-pathogen interactions; when rice cultivars were exposed to elevated CO_2 (approximately 100–300 μmol mol^{-1} higher than ambient) in open-top chamber, the disease was more severe under high CO_2 concentration and area under disease progress curve was 35.43 under high CO_2 concentration and 17.48 for the normal concentration. Elevated CO_2 levels did not alter the occurrence of foliage-infecting pathogens, viz. *M. oryzae*, *Bipolaris oryzae*, *Phoma sorghina*, *Drechslera* spp., *Alternaria* spp. or *Microdochium oryzae*. Moreover, leaves of treated rice plants with CO_2 which contain less silicon have showed that leaves were more prone to foliar diseases (Goria et al. 2013). Severity of blast and sheath blight is associated with reduced silicon content in susceptible rice cultivars under elevated CO_2 (Kobayashi et al. 2006). Tonkaz et al. (2010) reported that elevated CO_2 levels also positively affected yield, grain number, leaf area and biomass. However, elevated CO_2 levels reduced harvest index and evapotranspiration but did not had any effect on flowering date, maturity and 1000 seed weight. Rodrigues and Datnoff (2005) already reported that rice cultivars which contain less silicon were more prone to foliar disease infestation, especially blast and brown leaf spot diseases, than high silicon-containing rice cultivars (Datnoff et al. 1991; Rodrigues and Datnoff 2005).

8.5 Symptoms, Pathogenesis and Management

Rice blast is a major foliage disease problem in tropical and temperate regions and is distributed in irrigated, lowland and upland rice-producing areas. The favourable conditions for rice blast include long periods of free moisture where leaf wetness and high humidity are required for infection. Spore germination, infection and lesion formation are at optimum levels at 25–27 °C while sporulation occurs in high relative humidity and temperature (25–27 °C). Severity of blast and sheath blight is associated with reduced silicon content in the leaves of susceptible rice varieties (Kobayashi et al. 2006). Additionally, increased leaf wax and epidermal thickness in rice are greater influence of physical susceptibility to pathogens along with better pathogen fecundity and changes in pathogen virulence and distribution (Plessl et al. 2005). Moreover, Matros et al. (2006) reported that elevated CO_2 modifies secondary metabolites which influence pathogen (potato virus Y) ingress in tobacco.

8.5.1 Symptoms of Rice Blast

Disease symptoms are observed on all above ground parts of the rice plant. Blast pathogen produces lesions or spots on different parts of the rice plant such as leaf, leaf collar, panicle, culm and nodes. Initial symptoms are white to grey-green lesions/spots with darker borders produced on all infected shoots and leaves, while older lesions are elliptical/spindle shaped and whitish to grey with necrotic borders and these lesions may enlarge and coalesce to kill the entire leaf. Small specks originate on leaves—subsequently enlarge into spindle-shaped spots (0.5–1.5 cm length, 0.3–0.5 cm width) with ashy centre. Sometimes internodal infection of the culm gives banded pattern of lesions. Nodal infection causes the culm to break at the infected node (rotten neck); that is why the disease is popularly called 'rotten neck disease of rice'. As a result of the disease, the plant produces fewer seeds with dull and poor quality. Disease-causing pathogen can infect paddy at all stages of growth of rice from rice seedling to matured plants. It is well observed that the infection at three leaves and neck infections may cause severe yield loss than other stage of infection.

8.5.2 Pathogenesis

Disease-causing pathogens survive in the form of conidia on or inside the seed and perithecia on infected plant debris. Primary infection is caused by activated fungal mycelium or conidia and ascospores which germinate and cause primary infection when favourable environmental condition occurs (see Table 8.1) while secondary infection is caused by asexual spores, i.e. conidia (Fig. 8.1). The infection route requires an infection peg, called an appressorium, which uses a pressure-driven mechanism to break the tough cuticle of the rice plant and sticks firmly by means of an adhesive carried in the spore apex, generating turgor pressure of up to 8.0 MPa that ruptures the cuticle of the affected rice. Once inside the tissue, the fungus produces invasive hyphae that quickly colonise living host cells, secreting effector molecules to overpower host immunity and support infection. The effectors are transported into host cytoplasm by the aid of a biotrophic interfacial complex, a plant-derived membrane-rich structure in which effectors amass during transit to the host (Kankanala et al. 2007). The pathogen can replicate quickly and successively by mitosis, nuclear migration and death of conidia from which the infection originated, and produce appressoria capable of infecting aerial structures and hyphae capable of infecting roots of young and old rice plants. Autophagic cell death of conidia is connected to cell cycle control and produces conidiophores that are dispersed to other tissues and plants by wind and water splash to reinitiate the infection cycle by attachment of a spore that germinates and forms an appressorium. This allows the pathogen to infect epidermal cells with bulbous invasive hyphae that proliferate and grow from cell to cell, often through pit fields which invade neighbouring cells through

Table 8.1 Environmental factors which favour blast disease development in rice

Conditions	Stages	Range (°C)	Optimum (°C)
Leaf wetness	All stages	Always required	
Air temperature	Appressorium germination	10–33	25–28
	Appressorium formation	21–30	28
	Lesion formation (wet leaves)		4–5 days at 25–28
	Mycelium growth	8–37	28
	Mycelium survival for 18 months	−20 to −30	−30
	Sporulation	9–35	25–28
	Dispersal of conidia		20.5–21.8
	Host blast susceptibility	10–30	25–28
Soil temperature	Rice seedlings	20–30	
	Adult plants	18–24	
RH (air)	Mycelial growth	89–96	93%
	Conidial condition		93%
	Dispersal of conidia		90%
	Disease development		93–95%
Rainfall	All stages (direct effect)	Unclear	Unclear
Sunlight	Lesion formation		Night hours
Near-UV light	Germ tube length		
CO_2	Ambient +200–300 μmol mol^{-1}		

Source: Katsantonis et al. (2017) Phytopathologia Mediterranea (2017), 56, 2, 187–216, www.fupress.com/pm ISSN (print): 0031-9465 Firenze University Press ISSN (online): 1593-2095. DOI: https://doi.org/10.14601/Phytopathol_Mediterr-18706

plasmodesmata that requires mitogen-activated protein kinase signalling and manipulation of jasmonate signalling (Kankanala et al. 2007; Patkar et al. 2015). Appressorium penetration is a septin-dependent process and is linked to a burst of reactive oxygen species in the infected cell (Kankanala et al. 2007). Rice blast conidia can spread within 230 m from their source; dispersal is favoured in darkness and with high relative humidity and winds greater than 3.5 m s^{-1}. The primary source of inoculum is infected residue and seeds of rice, and in the tropics, airborne conidia are present throughout the year, enabling stable epidemics to occur year-round (Guerber and TeBeest 2006; Raveloson et al. 2018).

8.5.3 Management

Integrated disease management strategies are required for effective successful management of rice blast by including all the available options of disease control like physical, chemical and biological agents; selection of advanced breeding lines and cultivars with resistance genes; disease forecasting; and mapping distribution of the disease. These available tactics should be integrated with agronomic practices including the removal of crop residues to decrease pathogen survival, collateral

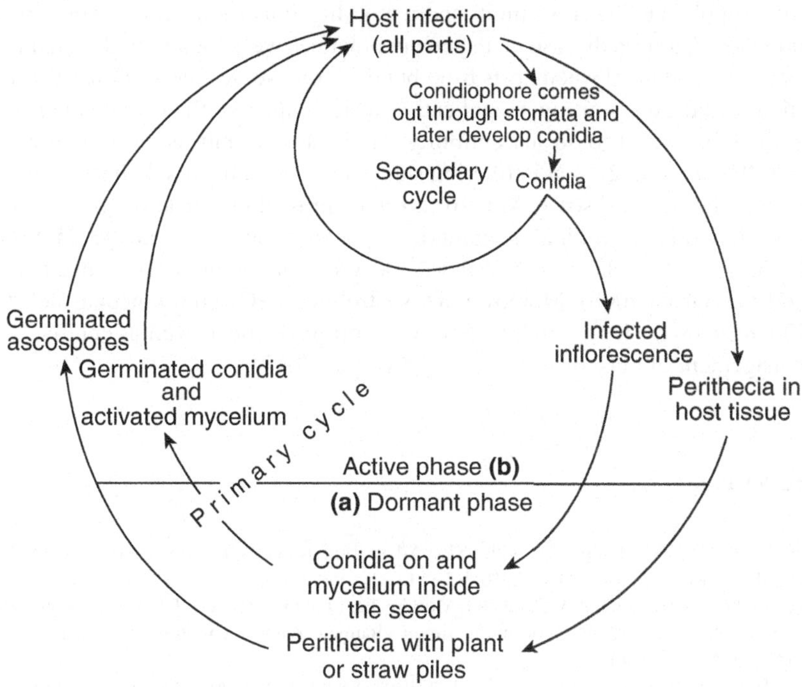

Fig. 8.1 Blast disease cycle (*Magnaporthe grisea*) (**a**). In dormant phase: in the form of conidia that survive on mycelium inside of the infected seeds or perithecia on infested plant debris or residues. (**b**). Active phase: dormant activated mycelia or conidia and ascospores germinate and cause primary infection while secondary infection causes conidia

host, adoption of crop and land rotations, avoiding of broadcast planting and double cropping, water management and balanced nutrient management (Asibi et al. 2019). Excessive use of nitrogen fertilisation as well as drought stress increases rice susceptibility to foliar disease-causing pathogens which leads to plant placed in a weakened position and its defences in weaker zone. There are two basic techniques that can be adopted for successful management of blast with the chemical fungicide strategy. In the first technique, seed treatment is used to prevent infection in seedlings after germination while in the second technique, fungicides are used to prevent infection of leaves and panicles during the growing season by making one or two foliar applications of fungicides to protect the panicles when they are emerging from the boot. This technique attempts to reduce the incidence of rice blast of seedlings, panicle necks and panicles. The most efficient way to control infection by *M. oryzae* is adopting of integrated disease management approaches. For example, eliminating crop residue could reduce the occurrence of overwintering and discourage inoculum load in subsequent seasons. Use resistant rice varieties to minimise yield losses. Knowledge of the pathogenicity of *M. grisea* and its need for free moisture suggests other control strategies such as regulated irrigation and a combination of chemical treatments with different modes of action. Managing the amount of water supplied to the crops limits spore mobility, thus dampening the opportunity for infection. Cultural disease management practices were found highly satisfactory by removing collateral weed hosts from bunds. Use disease-free seedlings for transplanting; avoid excess nitrogen application which enhances the leaf area in per unit area which induces disease susceptibility. Application of nitrogen (N) in three split doses (50% N basal, 25% N in tillering stage and 25% N in panicle initiation stage) minimises the risk of disease. Moreover, foliar spray of chemical fungicides immediately after disease initiation/symptoms appears with tebuconazole 75 WG @ 500–750 g/ha, tricyclazole 75 WP @ 500 g/ha, or metominostrobin 20 SC @ 500 mL/ha (Groth 2006). Moreover, azoxystrobin 25 SC and propiconazole 25 EC @ 500 mL/ha were found highly effective as prophylactic as well as curative mode of management of blast disease of rice (Pak et al. 2017).

References

Aaron N, Brooks ST, Fang YL, Nitin SB. Adaptation of cells to new environments. Wiley Interdiscip Rev Syst Biol Med. 2010;3(5):544–61. https://doi.org/10.1002/wsbm.136.

Anderson PK, Cunningham AA, Patel NG, Morales FJ, Epstein PR, Daszak P. Emerging infectious diseases of plants. Pathogen pollution, climate change and agrotechnology drivers. Trends Ecol Evol. 2004;19:535–44.

Asibi AE, Qiang C, Coulter JA. Rice blast: a disease with implications for global food security. Agronomy. 2019;9:451. https://doi.org/10.3390/agronomy9080451.

Baker JT. Yield responses of southern US rice cultivars to CO2 and temperature. Agric Forest Meteorol. 2004;122:129–37.

Beaumont HJ, Gallie J, Kost C, Ferguson GC, Rainey PB. Experimental evolution of bet hedging. Nature. 2009;462(7269):90–3.

Challinor AJ, Watson J, Lobell DB, Howden SM, Smith DR, Chhetri N. A meta-analysis of crop yield under climate change and adaptation. Nat Clim Change. 2014:287–91. https://doi.org/10.1038/nclimate2153.

Chesson P. General theory of competitive coexistence in spatially-varying environments. Theor Popul Biol. 2000;58:211–37.

Couch BC, Kohn LM. A multilocus gene genealogy concordant with host preference indicates segregation of a new species, *Magnaporthe oryzae*, from *M. grisea*. Mycologia. 2002;94:683–93.

Datnoff LE, Snyder GH, Raid RN, Jones DB. Effect of calcium silicate on blast and brown spot intensities and yields of rice. Plant Dis. 1991;75:729–32.

Godfray HCJ, Beddington JR, Crute IR, Haddad L, Lawrence D, Muir JF, Pretty J, Robinson S, Thomas SM, Toulmin C. Food security: the challenge of feeding 9 billion people. Science. 2010;327:812–8.

Goria MM, Ghini R, Bettiol W. Elevated atmospheric CO2 concentration increases rice blast severity. Trop Plant Pathol. 2013;38(3):253–7.

Groth DE. Azoxystrobin rate and timing effects on rice head blast incidence and rice grain and milling yields. Plant Dis. 2006;90:1055–8.

Guerber C, TeBeest DO. Infection of rice seed grown in Arkansas by (*Pyricularia grisea*) and transmission to seedlings in the field. Plant Dis. 2006;90:170–6.

Hughes BS, Cullum AJ, Bennett AF. An experimental evolutionary study on adaptation to temporally fluctuating pH in *Escherichia coli*. Physiol Biochem Zool. 2007;80:406–21.

Kankanala P, Czymmek K, Valent B. Roles for rice membrane dynamics and plasmodesmata during biotrophic invasion by the blast fungus. Plant Cell. 2007;19:706–24.

Kassen R. Toward a general theory of adaptive radiation. Ann N Y Acad Sci. 2009;1168(1):3–22.

Katsantonis D, Kadoglidou K, Puigdollers P. Rice blast forecasting models and their practical value. Phytopathol Mediterr. 2017;56:187–216.

Kobayashi T, Ishiguro K, Nakajima T, Kim HY, Okada M, Kobayashi K. Effects of elevated atmospheric CO2 concentration on the infection of rice blast and sheath blight. Phytopathology. 2006;96:425–31.

Matros A, Amme S, Kettig B, Buck SGH, Sonnewald UWE, Mock HP. Growth at elevated CO2 concentrations leads to modified profiles of secondary metabolites in tobacco cv. Samsun NN and to increased resistance against infection with potato virus Y. Plant Cell Environ. 2006;29:126–37.

McDonald BA, Linde C. Pathogen population genetics, evolutionary potential and durable resistance. Annu Rev Phytopathol. 2002;40:349–79.

Mitchell A, Romano GH, Groisman B, Yona A, Dekel E, Kupiec M, Dahan O, Pilpel Y. Adaptive prediction of environmental changes by microorganisms. Nature. 2009;460:220–4.

Padmanabhan SY. Studies on forecasting outbreaks of blast disease of rice. Proc Indian Acad Sci Sect B. 1965;62:117–29.

Pak D, You MP, Lanoiselet V. Azoxystrobin and propiconazole offer significant potential for rice blast (*Pyricularia oryzae*) management in Australia. Eur J Plant Pathol. 2017;148:247–59.

Patkar RN, Benke PI, Qu Z, Chen YYC, Yang F, Swarup S, Naqvi NI. A fungal monooxygenase-derived jasmonate attenuates host innate immunity. Nat Chem Biol. 2015;11:733.

Plessl M, Heller W, Payer HD, Elstner EF, Habermeyer J, Heiser I. Growth parameters and resistance against Swedish net blotch (*Drechslera teres*) of spring barley (*Hordeum vulgare* L. cv. Scarlett) grown at elevated ozone and carbon dioxide concentrations. Plant Biol. 2005;7:694–705.

Porter JR, Xie L, Challinor AJ, Cochrane K, Howden SM, Iqbal MM, Lobell DB, Travasso MI, Netra CNC, Garrett K, Ingram J, Lipper L, McCarthy N, McGrath J, Smith D, Thornton P, Watson J, Ziska L. Food security and food production systems. 2014. p. 485–533.

Raveloson H, Ratsimiala RI, Tharreau D, Sester M. Long term survival of blast pathogen in infected rice residues as major source of primary inoculum in high altitude upland ecology. Plant Pathol. 2018;67:610–8.

Rodrigues FA, Datnoff LE. Silicon and rice disease management. Fitopatol Bras. 2005;30:457–69.

Savary S, Mille B, Rolland B, Lucas P. Patterns and management of crop multiple pathosystems. Eur J Plant Pathol. 2006;115:123–38.

Savary S, Nelson A, Willocquet L, Pangga I, Aunario J. Modelling and map-ping potential epidemics of rice diseases globally. Crop Prot. 2012;34:6–17.

Teng PS. The epidemiological basis for blast management. In: Zeigler RS, Leong SA, Teng PS, editors. Rice blast disease. Wallingford: CAB International and IRRI; 1994. p. 409–33.

Teng PS, Klein-Gebbinck HW, Pinnschmidt H. An analysis of the blast pathosystem to guide modeling and forecasting, In: Rice Blast Modeling and Forecasting: Selected Papers from the International Rice Research Conference, 27–31 August 1990, Seoul, Korea. Edited by International Rice Research Institute, Manila, Philippines. 1991. p. 1–30.

Tonkaz T, Dogan E, Kocyigit R. Impact of temperature change and elevated carbon dioxide on winter wheat (*Triticum aestivum* L.) grown under semi-arid conditions. Bulgarian J Agric Sci. 2010;16(5):565–75.

Veeraraghavan J, Padmanabhan SY. Studies on the host range of *Pyricularia oryzae* Cav. causing blast disease of rice. Proc Indian Acad Sci Sect B. 1965;61(2):109–20.

Yang L, Huang J, Yang H, Dong G, Liu G, Zhu J, Wang Y. Seasonal changes in the effects of free-air CO2 enrichment (FACE) on dry matter production and distribution of rice (*Oryza sativa* L.). Field Crop Res. 2006;98:12–9.

Chapter 9
The Blast: A Major Malady in Nutricereals in Southeast Asia

T. S. S. K. Patro, K. E. Georgia, B. Praveen, K. B. Palanna, and N. Anuradha

Contents

9.1 Introduction

Small millets are traditional small-grained cereal food crops, comprised of finger millet (*Eleusine coracana*), foxtail millet (*Setaria italica*), little millet (*Panicum miliare*), barnyard millet (*Echinochloa frumentacea*), proso millet (*Panicum miliaceum*), kodo millet (*Paspalum scrobiculatum*) and browntop millet (*Panicum ramosa*). Most of the millets are native of India and are known as nutricereals as they provide most of the nutrients required for the normal functioning of a human body. Besides, they can withstand drought conditions and other extreme situations. Moreover, they can be grown with ease with less inputs such as fertilisers and pesticides. In addition to high-quality fodder crops, owing to their very high nutritive value, they have become popular in human diet. In India, minor millets' annual production is about 3.30 lakh tonnes.

All the small millet crops are known to be less prone to diseases. However, blast disease is one of the important yield-limiting constraints especially in finger millet

T. S. S. K. Patro (✉) · K. E. Georgia · N. Anuradha
Agricultural Research Station, ANGRAU, Gajularega, Vizianagaram, Andhra Pradesh, India

B. Praveen · K. B. Palanna
Project Coordinating Unit, ICAR-AICRP on Small millets, UAS, GKVK, Bengaluru, Karnataka, India

S. C. Nayaka et al. (eds.), *Blast Disease of Cereal Crops*, Fungal Biology, https://doi.org/10.1007/978-3-030-60585-8_9

and foxtail millet crops majorly grown in Andhra Pradesh, Haryana, Madhya Pradesh, Maharashtra and Mysore. The degree of damage depends on the extremity and time of onset of disease in finger millet. In finger millet, blast disease causes an average loss of about 28%, whereas in endemic areas it is to the extent of 80–90% (Patro and Anuradha 2019). Blast disease in finger millet is significantly the most devastating, causing over 50% yield loss. Mc Rae (1922) reported that blast disease causes yield loss of about 56%, while it was about 80% in Mysore (Venkatarayan 1947). Nagaraja et al. (2007) stated that the ultimate loss in grain yield is due to enhanced spikelet sterility and reduction in grain weight and number. Blast disease in its severe form causes up to 30–40% loss of yield in foxtail millet (Nagaraja et al. 2007).

9.2 Pathogen

Pyricularia grisea (Cke.) Sacc.—Finger millet

Pyricularia setariae Nishikado.—Foxtail millet

It has been reported to be as *P. eleusinis* in some records. Park (1932) reported *P. oryzae* on finger millet in Uganda. In Malaya also *P. oryzae* has been recorded on this host. The isolate from finger millet does not infect rice. Ramakrishnan (1948) recorded it to be a race of *P. oryzae*. Wallace and Wallace (1948) also reported *P. oryzae* from Tanzania on finger millet. Wallace (1950) classified it to be as *P. setariae*. Nishikado (1917) reported the fungus occurring on the foxtail millet as *P. setariae*. According to Ramakrishnan (1948) the fungus on foxtail millet should be considered as a strain of *P. oryzae*, and not as a separate species. However, *P. setariae* name is generally used by plant pathologists.

Young hyphae are hyaline and septate, and older hyphae are brown coloured. The length of cells varies from 1.5 to 6.0 μ. Under humid conditions the central portion of the lesion possesses several conidia and conidiophores. The upper surface is darker than lower surface. From the stomata or epidermal cells, the conidiophores emerge. They are septate, straight and sub-hyaline at the top and dark at the bottom. Conidia appear hyaline, thin walled and subpyriform. The size of the conidia varies, being 19–31 μ × 10–15 μ. Each spore is tri-celled, with the centre cell being darker and broader than others. They are formed acrogenously one after another; by the sympodial growth of the conidiophore only end cells form germ tube. On germination in culture media the fungus produces olive brown to dark brown, globose chlamydospores, 4–10 μ in diameter. These may be terminal or intercalary.

The pathogen grows on various media in the laboratory and produces a dark grey aerial growth. It grows well on the extracts of the host material. Nishikado (1927) reported that *Setaria* isolate grew best at a temperature of 23–28 °C. Optimum temperature for the growth of the finger millet isolate on culture media was 29.5 °C (Thomas 1940).

Maximum growth of finger millet isolate was obtained on solid media at 30 °C (Ramakrishnan 1948). There was no significant difference in mat weights between growths at 20 and 30 °C. The minimum and maximum temperatures for the fungus were 5 °C and 36 °C, respectively. According to Thomas (1940) the optimum reaction of the medium was between pH 5 and 6; Ramakrishnan (1948) recorded the best growth of the fungus at pH 7. The pH of the medium was changed to about 5 and 6, during the growth of the fungus. The spores die when exposed for 5 min at 48–49 °C. Finger millet isolate easily utilised different carbon sources. The best sources of carbon are fructose, mannose, sucrose and glucose. The nitrogen sources are inorganic nitrate nitrogen, organic amide and amino nitrogen. Suryanarayanan (1958) reported that ammonium oxalate, ammonium tartrate, urea, potassium nitrate and sodium nitrate are the better sources. Fungus produces several enzymes, *i.e.* lipase, inulinase, amidase, diastase, sucrose, erepsin, lactase and trypsin (Ramakrishnan 1948). Sadasivan and Subramanian (1954) reported that the fungus was heterotrophic for thiamine and biotin. Suryanarayanan (1958) also reported that the fungus is with thiamine and biotin deficiency. A study shows that pyridoxine, inositol and nicotinic acid slightly stimulated growth. The neck isolate responded more to vitamins than the leaf blast isolates (Kulkarni and Govindu 1976).

9.3 Symptoms

Finger millet is susceptible from seedling stage to the grain formation stage. Young seedlings are infected in the nursery and in the fields, with spindle-shaped lesions of various sizes. The spot begins as yellowish margin with greyish green centre and later the center becomes whitish grey and disintegrates. Under humid conditions, an olive grey overgrowth of fungus develops, at the centre of the spots. Conidiophores and conidia are present in the overgrowth. The distal portions of the leaves beyond lesions may hang and drop off. In the beginning the lesions are isolated and afterwards they may soon coalesce.

Lesions develop on leaf blades of the adult plants. The lesions are like those on the seedlings and are about 0.3–1.0 cm in breadth and about 1–2 cm in length. The apices of the infected leaves beyond the lesions hang down and sometimes break. Nodal regions of the stem blacken and penetrate into the tissues due to infection. Neck infection causes maximum damage and region of the neck shrinks and becomes black. This area is seen with growth of olive grey fungus. The ear hangs down from the stalk and sometimes it may break away. Infections result in browning of the basal regions of the panicle along with fingers. Infected ears show shrivelled and chaffy grains, and the effected portions of the ear head become black. The loss in grain yield depends upon the time of infection.

Leaf blade shows spots in foxtail millet. Spots are round with pale centre and are encircled with dark brown margin. They are minute and dispersed and measure a diameter of 1–5 mm severe infection, leading to drying and withering of leaves (Fig. 9.1).

Fig. 9.1 Symptoms of blast disease: (**a**) leaf blast; (**b**) neck blast; (**c**) finger blast; and (**d**) Setaria leaf blast

9.4 Infection

Fungus enters the host by piercing through the epidermal cells or through stomata. The incubation period varies from 4 to 6 days. Based on weather conditions the intensity of disease varies accordingly. The crop which is sown in June–July is affected worst because they are exposed to high humidity due to monsoon rains. Continuous rains during ear head formation lead to heavy loss to the crop. Thomas (1940) reported that heavy infections of the neck and ear when sowings were done in June (69–81%). No infections were recorded in the crops sown in January to April or October to December, in the year 1941; 31% of infection was recorded in the October sowing.

Seedlings are more susceptible than mature plants. Between ear infection and seedling infection no correlation was seen. It depends on the climatic conditions prevalent at the particular stage. The initial inoculum comes from weeds, collateral hosts, plant debris and shrivelled seeds. Thomas (1940) reported that spores were found on shrivelled seeds. Kato et al. (1977) also recorded the fungus on seeds. The pathogen readily infects foxtail millet, bajra, finger millet, wheat, barley, oats, maize and crowfoot grass. Kato et al. (1977) reported that *Pyricularia* isolates from *E. coracana, E. indica, E. africana* and *E. floccifolia* were pathogenic to finger millet. Isolates of *Pyricularia* sp. from *E. coracana, E. indica, E. africana* and *E. floccifolia* were pathogenic to ragi and not pathogenic to rice.

9.5 Management

9.5.1 Finger Millet Blast

The fungus is seed borne; seed treatment with organic mercurials reduces the incidence of disease in the nursery. At the time of transplanting, the infected leaves are clipped off and the seedlings are dipped in Bordeaux mixture. The secondary infection is through airborne conidia. It can be checked by spraying Bordeaux mixture. Deshkar et al. (1973) reported that Benlate is found effective in reducing neck infection. Shivaprakasam et al. (1974) found Ceresan lime dust to be the best for controlling the disease and it increased the yield by 36.8% over control. Shivaprakasam and Pillayarsamy (1975) recommended miltox and zineb for the control of finger millet blast. Carbendazim or *Pseudomonas fluorescens* as seed treatment showed decreased blast disease incidence (Nagaraja et al. 2012).

Deshkar et al. (1973) reported varieties PR 722 and PR202 as relatively resistant. Two cultivars, *viz.* TAH 14-8 and TAH 91-1, are relatively resistant (Pall and Nema 1978). Screening of 32 finger millet genotypes against leaf, neck and finger blast showed that WN 259, DHFMV78-3-1 and KRI 013-11 were relatively resistant to neck blast and susceptible to finger blast (Patro and Madhuri 2014). Neeraja et al. (2015) revealed that from 25 finger millet genotypes KMR 502, GPU 67, DHWFM 11-3, KOPN 930, TNEC 1277, TNEC 1269, GPU 45, PRSW 43 and DHFM 103 were relatively resistant/relatively susceptible against leaf blast. PR 10-51, KMR 346 and DHFM 103 were moderately resistant/moderately susceptible genotypes against finger and neck blast. Patro et al. (2016) identified that DHFMV 10-2-1 was resistant to blast disease when evaluated against seven varieties of finger millet. Nine finger millet genotypes were found resistant to blast disease by Patro et al. (2018a, b) depending on the collected data at all the centres, and revealed that least percentage of finger and neck blast was noticed in KOPN 942 (2.91) and PR 202 (12.05).

Finger millet genotypes are assessed for resistance; out of 30 varieties 5 of them are highly resistance to leaf blast, *viz.* PR 1507, WN 585, OEB 602, IIMR FM 6655 and GMB. Genotype (WN 550) recorded minimum incidence of neck and finger blast disease with 13.67% and 11.58%, respectively (Patro et al. 2018a, b). Patro et al. (2019a, b) assessed 3000 finger millet lines against finger and neck blast during *Kharif*, 2013–2018; among all the 3000 lines 112 lines showed resistant reaction under high-disease-pressure field conditions during *Kharif* 2013. However, only 50 varieties have shown consistent reaction during all the years (2014–2018). The minimum neck incidence was recorded in VR 1062 and VR 1104 with 2.5% and VR 1080 recorded 0.0% of finger blast.

Twenty-six varieties of finger millets were evaluated against blast disease by Patro et al. (2019a, b) found that KMR 650 and RAuF 17 were highly resistant to leaf blast. KMR 650 recorded minimum per cent of neck and finger blast incidence with 13.2 and 13.0, respectively. Patro et al. (2019a, b) revealed that VR 1101 (16.5 and 18.6%) showed minimum percentage of finger and neck blast incidence among the 19 finger millet genotypes. In vivo experiments were conducted by Patro and Georgia (2020) against finger millet blast with nine treatments, viz. seed treatment with carbendazim, chitosan, *P. fluorescens* and *Bacillus subtilis*; seed treatment with carbendazim + two foliar sprays with *P. fluorescens*; seed treatment with chitosan + two foliar sprays with *P. fluorescens*; seed treatment with *P. fluorescens* and two foliar sprays with *P. fluorescens*; and seed treatment with *B. subtilis* + two foliar sprays with *B. subtilis* and untreated check. Among these nine treatments seed treatment with chitosan + two foliar sprays with *P. fluorescens* was superior in managing blast disease with lowest incidence of neck and finger blast 8.7% and 7.7%, respectively, when compared to untreated check (80.0 and 77.7) percent.

9.5.2 Foxtail Millet Blast

Goel et al. (1967) reported that *P. setariae* was found on 3–4% of the seeds of the variety ISc470; the first report of this fungus was as seed-borne pathogen. Mathur et al. (1967) reported that Agrosan 5W was the most effective in controlling seed-borne infection of blast. Varieties ISc709, ISc701, SR 118, RS179, ISc703, ST5307, ISc710, JNSe33, SR102, JNSe56, ISc201, highly resistant ISc358, ISc700, Co3, Arjuna, ISc709, ISc480, JNse 15A, JNSe9A, ST 8012 and ST 7614 were found resistant (Singh et al. 1976). Thirteen varieties of foxtail millet were evaluated against blast disease; among these two varieties, i.e. SiA-2679 and SiA-2676, were found resistant. Nagaraja et al. (2007) reported that edifenphos @ 1 mL/L, carbendazim @ 1 g/L, or carbendazim + mancozeb @ 1 g/L was found superior in managing blast disease in foxtail millet. Sharma et al. (2014) assessed 154 accessions against *P. setariae* and found that four accessions ISc1181, ISc1547, ISc1067 and ISc1575 showed blast resistance.

References

Deshkar MV, Sharma BL, Dhagat NK, Joshi RC. Varietal susceptibility to and evaluation of fungicides against blast of finger millet. JNKVV Res J. 1973;7(4):295–7.

Goel LB, Mathur SB, Joshi LM. Seed borne infection of *Pyricularia setariae in Setaria italica*. Plant Dis Rep. 1967;5(2):138.

Kato H, Yamaguchi T, Nishihara N. Seed transmission, pathogenicity and control of finger millet blast fungus and susceptibility of finger millet to *Pyricularia* spp. from grasses, cereals and mioga. Ann Phytopathol Soc Jpn. 1977;43(4):392–401.

Kulkarni S, Govindu HC. Studies on the blast disease of finger millet on Karnataka. II Vitamin requirement soft he leaf and neck isolates of *Pyricularia setariae* Nishikado. Mysore J Agric Sci. 1976;10(4):627–31.

Mathur SB, Goel LB, Joshi LM. Fungi associated with seeds of *Setaria italic* and their control. Proc Int Seed Test Assur. 1967;32(3):633–8.

MC RAE, W., 1922. Report of the Imperial Mycologist. Sci. Rep. agric. Res. Inst, Pusa (1921–22). pp. 44–50.

Nagaraja A, Kumar J, Jain AK, Narasimhudu Y, Raghuchander T, Kumar B, Gowda BH. Compendium of small millets diseases. Bangalore: AICSMIP; 2007.

Nagaraja A, Kumar B, Raghuchander T, Hota AK, Patro TSSK, Gowda PD, Savita E, Gowda MVC. Impact of disease management practices on finger millet blast and grain yield. Indian Phytopathol. 2012;65(4):356–9.

Neeraja B, Patro TSSK, Sandhya RY, Jyothsna S. Studies on three forms of blast (leaf, neck and finger) in finger millet (*Eleusine coracana* Gaertn.) incited by *Magnaporthe grisea* (Hebert). Barr. under *in vivo*. Prog Res Int J. 2015;10(4):2098–100.

Nishikado Y. Studies in the rice blast fungus. I. Ber Ohara Inst. 1917;1:172–219.

Nishikado Y. Studies in the rich blast fungus. Jpn Bot. 1927;3:239–44.

Pall BS, Nema AG. Screening of early varieties of finger millet against blast disease. Food Farming Agriculture. 1978;X:N.5.86.

Park M. Report on the work of the mycological division. Adm Rep Dir Agric Ceylon 1931. 1932. p. 103–11.

Patro TSSK, Anuradha N. Identification of resistant sources of finger millet lines against neck and finger blast disease. Int J Chem Stud. 2019;7(2):3601–4.
Patro TSSK, Georgia KE. Annual reports. Agriculture Research Station, Vizinagaram. 2020.
Patro TSSK, Madhuri J. Identification of resistant varieties of finger millet for leaf, neck and finger blast. Int J Food Agric Vet Sci. 2014;4(2):7–11.
Patro TSSK, Divya M, Sandhya RY, Triveni U, Ashok S. Identification of resistant sources of finger millet against *Pyricularia grisea* Sacc. J Agric Technol. 2016;12(7):1519–23.
Patro TSSK, Meena A, Divya M, Anuradha N. Evaluation of finger millet early and medium duration varieties against major diseases. Int J Chem Stud. 2018a;6(3):2184–6.
Patro TSSK, Meena A, Divya M, Anuradha N. Screening of finger millet for major diseases and identification of resistant varieties. J Pharmacogn Phytochem. 2018b;7(3):2681–3.
Patro TSSK, Raj KS, Meena A, Anuradha N, Triveni U, Joga RP. Screening for identification of finger millet early and medium duration varieties against major diseases. Int J Chem Stud. 2019a;7(6):954–6.
Patro TSSK, Raj KS, Meena A, Anuradha N, Triveni U, Joga RP. Screening of finger millet varieties for major diseases and identification of resistant varieties. Int J Chem Stud. 2019b;7(6):1446–8.
Ramakrishnan KV. Studies on the morphology, physiology and parasitism of the genus *Pyricularia* Madras. Proc Indian Acad Sci B. 1948;27:174–93.
Sadasivan TS, Subramanian CV. Studies on the growth requirements of Indian fungi. Trans Br Mycol Soc. 1954;37:426–30.
Sharma R, Girish AG, Upadhyaya HD, Humayun P, Babu JK, Rao VP, Thakur RP. Identification of blast resistance in a core collection of foxtail millet germplasm. Plant Dis. 2014;98:519–24.
Shivaprakasam K, Pillayarsamy K. Efficacy of some chemicals in the control of finger millet blast. Madras Agric J. 1975;162(2):84–6.
Shivaprakasam K, Pillayarsamy K, Iyemperumal S, Ahmed NJ. Note on the effect of fungicides on the control of the finger millet blast. Indian J Agric Sci. 1974;44(4):249–50.
Singh BP, Pall BS, Nema AG, Deshakar MV. Varietal reaction to blast of Kangni. Curr Res I. 1976;13.
Suryanarayanan S. Growth factor requirement of *Pyricularia* spp. and *Sclerotium oryzae*. Indian Acad Sci B. 1958;58:154–88.
Thomas KM. Detailed administration report of the govt. Mycologist Madras for the year 1939-40. 1940. p. 18.
Venkatarayan SV. Diseases of finger millet *(Eleusine coracana)*. Mysore Agric J. 1947;24:50–7.
Wallace GB. Plant Pathology. Rep Agric Tanzania, 1948-45. 1950.
Wallace GB, Wallace M. Tanzania territory fungus list. Recent records. XI Mycol Cir Dep Agric Tanzania. 1948;24–25.

Chapter 10
Microconidia: Understanding Its Role in the Fungus *Magnaporthe oryzae* Inciting Rice Blast Disease

Ganesan Prakash, Asharani Patel, Ish Prakash, Kuleshwar Prasad Sahu, Rajashekara Hosahatti, and Aundy Kumar

Contents

10.1 Introduction

Globally, *Magnaporthe* species (syn. *Pyricularia* sp.) is a rapidly evolving pathogen that infects multiple grasses and cereal crops such as rice, pearl millet, finger millet, foxtail millet, wheat, oats and barley due to its adaptation to the new host via host shift/host jump. Consequently, it destroys food supplies that could feed hundreds of millions of people of the world (Fisher et al. 2012). Rice blast is one of the most destructive diseases caused by the fungus *Magnaporthe oryzae* in several parts of the world (Helliwell et al. 2013). Rice blast disease is a major biotic factor that contributes to almost more than 30% yield loss worldwide (Skamnioti and Gurr 2009; Nalley et al. 2016). It is observed that even moderate *M. oryzae* infections can cause approximately 50% grain yield reductions (Katsantonis et al. 2017). In India, the rice disease dynamics on yield loss shows that major yield loss of 35% due to

G. Prakash (✉) · A. Patel · I. Prakash · K. P. Sahu · A. Kumar
Division of Plant Pathology, ICAR-Indian Agricultural Research Institute, New Delhi, India

R. Hosahatti
ICAR-Vivekananda Institute of Hill Agriculture, Almora, India

S. C. Nayaka et al. (eds.), *Blast Disease of Cereal Crops*, Fungal Biology,
https://doi.org/10.1007/978-3-030-60585-8_10

rice blast, sheath blight (25%), bacterial leaf blight (20%), tungro virus (10%) and other diseases (10%) indicates the impact of rice blast pathogen (Kumar et al. 2013).

Rice blast is one of the most devastating rice diseases, affects grain quality and causes huge crop loss ranging from 50 to 80% depending on the severity and timing of infection. Rice blast is an attack by the fungus *M. oryzae* that colonises in leaves (leaf blast), panicles (panicle blast) and other parts of the rice plants in rice-growing areas. In heavy infections, disease symptoms can be so severe that whole seedlings may die, whereas in mature plants, the fungus can prevent grain-filling or destroy grain-bearing structures of the plant (Howard and Valent 1996; Tucker and Talbot 2001). Ascomycete plant pathogenic fungi produce more than one type of asexual spore in the anamorphic (asexual) stage, in which *M. oryzae* is not exceptional. Rice blast pathogen produces macroconidia and microconidia. In teleomorph (sexual) stage it produces four-celled, spindle-shaped ascospores (Chuma et al. 2009). The pathogenicity mechanism in rice blast fungus *M. oryzae* mainly involves asexual spores, namely macroconidia. So far, the studies on this fungus have reported macroconidia as the only source of infection in causing blast symptoms. But recently, it has been proven that microconidia can also cause infection on wounded rice plants, suggesting that macroconidia may not be the only source of inoculum in nature (Zhang et al. 2014).

10.2 Pathogen Biology

10.2.1 Sexual Reproduction

The perfect state of blast pathogen was first reported by Hebert (1971) by crossing two isolates of crabgrass host (*Digitaria sanguinalis*) from the southern United States. He named as *Ceratosphaeria grisea* for the perfect stage of blast pathogen and observed hermaphroditic and heterothallic ascomycete nature of the pathogen. Later in 1977, Barr, based on the perithecial morphological characteristics, imperfect stage of the pathogen and graminicolous habit, raised objection for placing it under the genus of *Ceratosphaeria* and renamed it as *Magnaporthe grisea*. After Hebert's interesting observation on the perfect stage formation of blast pathogen, there were numerous studies initiated throughout the world to understand about blast pathogen's sexual stage existence. As a result, perithecia from intragroup mating of rice isolates were first observed by Kato and Yamaguchi (1982). The key identification helped others to search for rice isolates that could be crossed.

Throughout the world, hermaphroditic strains were observed in rice blast pathogen (Li et al. 1992, 1996; Mekwatanakarn et al. 1999). Opposite mating types of *M. oryzae* pathogen (MAT1-1 and MAT1-2) produce ascospores in structures called asci that are found within specialised structures called perithecia as a sexual or teleomorphic stage. Unitunicate asci contain hyaline, fusiform-shaped (spindle shaped with tapering ends), three-septate ascospores (Zeigler et al. 1994). Single or a group of perithecia are formed, erumpent with long protruding beaks, dark brown to black, 60–300 μm in diameter. Unitunicate asci, cylindrical to clavate with

55–90 × 7–10 μm in size, contain three-septate ascospores, fusiform and hyaline in nature (Gupta and Kapoor 2002). This ascomycete fungus is now under a separate family Magnaportheceae under revised new taxonomy classification. Sexual reproduction occurs between these two opposite mating types; at least one of these which is female-fertile that comes into contact was observed under laboratory condition. In field condition, *Magnaporthe oryzae* may be predicted to reproduce sexually in some of the regions (Yunnan Province of China, Northern Thailand and Himalayan foothills of India) (Zeigler 1998; Mekwatanakarn et al. 1999; Kumar et al. 1999), based on the observation of hermaphroditic strains. The overall literature suggested that *M. grisea* from non-rice hosts exhibits high level of sexual fertility (Yaegashi and Nishihara 1976) and high degree of sexual compatibility among the rice- and non-rice-infecting isolates of *M. oryzae/M. grisea* (Rathour et al. 2004). So far, the blast pathogens infecting rice, wheat, turf grass and finger millet crops were reported at sexual (teleomorph) stage from the entire world (Consolo et al. 2005; Le et al. 2010; Marciel et al. 2014; Takan et al. 2012; Tredway et al. 2003; Saleh et al. 2012, 2014). Sexual recombination could be one of the reasons for its high variability and existence of new strains in nature which helps for host shift.

10.2.2 Asexual Reproduction

Pyricularia oryzae is the name for asexual stage of *Magnaporthe oryzae*. The mycelium of *M. oryzae* is septate, branched and mostly uninucleate hyphae. Conidiophores occur singly or in fascicles and are branched simply sometimes, most of the other times being unbranched. They are slender, septate, denticulate and greyish in colour exhibiting sympodial growth (Singh 2017). The tips of conidiophores bear conidia which are produced in succession, one at a time. It produces two types of asexual conidia, macroconidia and microconidia.

10.2.2.1 Macroconidia

Macroconidia (simply called as conidia) is the most common spore form of the fungus in the majority of the rice-growing countries. Conidia are two septate, ovate, tapering at the apex, 20–22 × 10–12 μm in size and borne terminally. Conidial shape is obclavate to narrow pyriform (pear shaped), its base being rounded, and narrows down towards the tip which is blunt or pointed depending on the fungal race. The conidium is two septate. Conidial cells are uninucleate and the nucleus has two large and two small chromosomes. Sometimes, rarely the conidia are one or three septate. A protruding hilum lies at the base. Each conidiophore may bear 20 or more number of conidia. 4000–6000 conidia are produced on a typical leaf lesion each night for 2 weeks or more. Conidiophores release conidia by dew or rain which then get disseminated by wind currents. Light influences the spore release behaviour in that even very dim light is capable of suppressing the release of spores. From their source of origin, most conidia travel only 1–2 m before landing on other rice plants or other leaves. The conidia germinate by forming several germ tubes in the presence of free water and favourable temperature within 3–4 h. Lipophilic self-inhibitors are known to be carried by conidia to prevent germination at the sporulation site. This self-inhibition can be relieved by the leaf cuticular wax when conidia land on the surface of the plant. Dome-shaped densely melanised appressoria are formed by the germ tubes. Appressorium formation is induced by some environmental factors like hardness thigmotropism of the contact surface, hydrophobicity and some host-produced chemicals. Virulence is related with the melanisation of appressoria. Mutants with melanin-deficient trait cannot effect proper penetration. Appressoria give rise to infection pegs which penetrate the tissues of the host. 7–8-h time is needed for accomplishing germination and penetration. About 4 days after spore germination, the lesions appear and in 6–7 days, a new conidial crop is produced (Hebert 1971; Ueyama and Tsuda 1975; Yaegashi and Nishihara 1976; Yaegashi and Udagawa 1978).

10.2.2.2 Microconidia

Kato et al. (1994) first time identified the unusual structures formed by *M. oryzae* and they named it as microconidia. They also noticed the phialides which are microconidial spore-bearing structures. They are thick-walled, dark pigmented structures, spherical to obclavate, with tapered apex, terminal collarette and basal septum emerging from aerial hyphae.

Microconidia are uninucleate, non-septate (whereas macroconidia two septate) hyaline, lunate with thin cell wall and size also smaller than macroconidia. During conidiogenesis process, conidiogenous cell apex bears two successive rod-shaped microconidia blastically from alternate side. Similar way more numbers of microconidia were produced from the same locus. Mass of microconidia are accumulated at the tip of phialide. Microconidia continue to grow and then secedes. Mature microconidia become crescent shape (Kato et al. 1994). Phialide structures are also reported from other species of Magnaporthe which are *M. rhizophila* (Cole and Samson 1979) and *M. poae* (Landschoot and Jackson 1989). In both the cases, the size and shape of the conidium are different from *M. oryzae*. Phialide and microconidial structures are also presented in other *Magnaporthe* species indicating the phylogenetic similarity among these species.

10.3 Cellular Structures of Microconidia

Under various microscopy observations, Chuma et al. (2009) observed that microconidia were having a larger portion of nuclei in their cells which are ellipsoidal in nature along the cell walls and also nucleus ratio to cell was the highest in microconidia. Thin cell walls and nucleoli were absent in microconidia when compared to macroconidia and vegetative hyphae whereas these structures had rice cytoplasm which contains a nucleus with one nucleolus, few mitochondria, endoplasmic reticulum and vacuoles. Cell size is small and nucleus size is also more or less similar or bigger in case of microconidia when compared to macroconidia and vegetative hyphae. Phialide nuclei are larger in size compared to any other nuclei (Chuma et al. 2009).

10.4 Biological Role of Microconidia

Under phylum Ascomycota, two major classes of fungi (Sordariomycetes and Leotiomycetes) produce microconidia. In Sordariomycetes, we have major orders producing microconidia, that is, Magnaporthales (e.g. *Magnaporthe*, *Gaeumannomyces*), Hypocreales (e.g. *Fusarium*) and Sordariales (e.g. *Neurospora*). In Leotiomycetes, we have an order Helotiales (e.g. *Botrytis*). In some cases the microconidia act as a spermatia in sexual reproduction (*Neurospora crassa*, *Botrytis cinerea* and *Podospora anserina*) (Fukumori et al. 2004; Maheshwari 1999). In these fungi microconidia nuclei occupy a major portion of the cell similar to *Magnaporthe* microconidia (Fukumori et al. 2004; Lowry et al. 1967; Zickler et al. 1995). When mating between opposite mating types (MAT1-1 and MAT1-2), abundant microconidia were produced, indicating that microconidia in *Magnaporthe* may play a role in fertility (Chuma et al. 2009). But later it has been proved that irrespective of the mating types, the microconidia are produced under submerged liquid cultures, and it was also predicted that microconidia may be playing a role in other biological functions such as decimation and disease spreading (Zhang et al. 2014).

10.5 Germination and Infection Process of Microconidia for Disease Establishment

In some of the fungal species, microconidia do not germinate (e.g. *Podospora*) (Esser 1974). In some cases it germinates naturally similar to macroconidia (e.g. *Fusarium oxysporum* and *F. verticillioides*) (Zhang et al. 2014) and germinates under certain specific nutrient supply (e.g. *Neurospora crassa*). In the case of *M. oryzae*, the microconidia germination percentage is very low (up to 10%) when compared to macroconidia on artificial and plant surfaces. Microconidia are able to germinate to form germ tube but fail to form appressorium. The pathogenicity mechanism in rice blast fungus *Magnaporthe oryzae* mainly involves asexual spores, namely macroconidia. So far, the studies on this fungus reported macroconidia as the only source of infection in causing blast symptoms. But recently, it has been proven that microconidia can also cause infection on wounded rice plants, suggesting that macroconidia may not be the only source of inoculum in nature.

Zhang et al. (2014) confirmed microconidial germination on plant surfaces up to 10% and observed normal growth and pathogenesis from the colonies derived from germinated microconidia. They proved that microconidia fail to infect intact plants in rice and barley seedlings, but it infects the wounded rice leaves and stems and produces typical necrotic lesions. Further, microconidia also show disease symptoms on inoculated spikelets in infection assays with barley and Brachypodium heads. Finally, they detected microconidia within rice plants that developed blast lesions under laboratory or field conditions. Ultimately, they proved that microconidia are capable of germination and cause disease symptoms in rice. Even though microconidia are able to germinate and form germ tube, it does not produce appressorium. So obviously direct penetration is not possible; that is why it gives symptom on wounded rice plants. On the other hand, microconidia are able to cause blast symptoms on barley and Brachypodium heads without wounding. It indicates that flowering stages are highly susceptible for microconidial infection.

Several reports suggested that *M. oryzae* can enter roots and spread without causing any external symptoms for quite a long period because of its hemibiotrophic nature (Dufresne and Osbourn 2001; Sesma and Osbourn 2004; Marcel et al. 2010). Ability of *Magnaporthe oryzae* to infect rice plants through roots in a systemic manner and also its prolonged period in biotrophic phase are reported. However, the importance of this type of infection in field epidemics still remains unclear (Ribot et al. 2008). The mode of infection in case of blast fungus is caused by aerial means and through root infection. In case of aerial infection biotrophic and necrotrophic growth is involved in expressing symptoms in aerial parts by forming appressoria. In case of root infection, it has been reported that prolonged biotrophic condition is maintained inside the plant and in later course blast symptom is expressed in aerial parts by forming hyphopodia rather than forming appressoria (Sesma and Osbourn 2004; Marcel et al. 2010). It is concluded that the ability to use either hyphopodia or appressorium thus permits *M. oryzae* to adapt its penetration mechanisms to the properties of the target organ, which in turn raises questions as to what extent similarity exists between leaf and root infection strategies and also what happens if biotrophic growth does not switch to necrotrophic growth at aerial parts but at neck zone which may lead to neck blast.

Moreover, under submerged condition, macroconidia are not able to form. Only microconidia are able to form. Systemic infection of *M. oryzae* indicates that may be microconidia form inside the plant tissues and it expresses symptoms under a particular stage that may be during panicle emergence for panicle blast disease establishment or microconidia may be involved in some other processes during pathogenesis. It was contemplated to associate the above-mentioned supposition with neck blast infection.

10.6 Conclusion and Future Perspectives

More than 100 years of research was carried out on rice blast pathogen. Recently we encountered microconidia as a novel structure in this pathogen. Rice blast pathogen *M. oryzae* produces macroconidia and microconidia as asexual spores. We have a better understanding about the macroconidial infection process, pathogenesis mechanisms and disease establishment. Presently we know that the uninucleate microconidia are also able to germinate independently irrespective of mating types and it can be able to cause blast symptoms in rice plants. In future, the biological role of microconidia should be established clearly. The pathogenesis mechanism of microconidia, relationship between microconidia and macroconidia and how it helps in disease establishment are not yet clear. How microconidia interact with the host, signal transaction between microconidia and host, its adaptation and survival strategy should be established. This will help us in better understanding of the rice blast pathogen which ultimately leads to devising of better management strategies.

References

Barr ME. *Magnaporthe, Telimenella,* and *Hyponeclaria (Physosporellaceae).* Mycologia. 1977;69:952.

Chuma I, Shinogi T, Hosogi N, Ikeda K, Nakayashiki H, Park P, Tosa Y. Cytological characteristics of microconidia of *Magnaporthe oryzae*. J Genet Plant Pathol. 2009;75:353–8.

Cole GT, Samson RA. Patterns of development in conidial fungi. London: Pitman; 1979. p. 54–72.

Consolo VF, Cordo CA, Salerno GL. Mating type distribution and fertility status in *Magnaporthe grisea* populations from Argentina. Mycopathologia. 2005;160:285–90.

Dufresne M, Osbourn AE. Definition of tissue-specific and general requirements for plant infection in a phytopathogenic fungus. Mol Plant Microbe Interact. 2001;14(3):300–7.

Esser K. Podospora anserina. In: King RC, editor. Handbook of genetics, vol. 1. New York: Plenum; 1974. p. 531–51.

Fisher MC, Henk DA, Briggs CJ, Brownstein JS, Madoff LC, McCraw SL, Gurr SJ. Emerging fungal threats to animal, plant and ecosystem health. Nature. 2012;484:186–94. https://doi.org/10.1038/nature10947.

Fukumori Y, Nakajima M, Akutsu K. Microconidia act the role as spermatia in the sexual reproduction of *Botrytis cinerea*. J Gen Plant Pathol. 2004;70:256–60.

Gupta VK, Kapoor AS. Diseases of field crops. New Delhi: Indus Publishing House; 2002.

Hebert TT. The perfect stage of *Pyricularia grisea*. Phytopathology. 1971;61(1):83–7.

Helliwell EE, Wang Q, Yang Y. Transgenic rice with inducible ethylene production exhibits broad spectrum disease resistance to the fungal pathogens *Magnaporthe oryzae* and *Rhizoctonia solani*. Plant Biotechnol J. 2013;11:33–42.

Howard RJ, Valent B. Breaking and entering–host penetration by the fungal rice blast pathogen *Magnaporthe grisea*. Annu Rev Microbiol. 1996;50:491–512.

Kato H, Yamaguchi T. The perfect state of *Pyricularia oryzae* cav. from rice plants in culture. Ann Phytopathol Soc Jpn. 1982;48:607–12.

Kato H, Mayama S, Sekine R, Kanazawa E, Izutani Y, Urashima AS, Kuhoh H. Microconidium formation in *Magnaporthe grisea*. Annu Phytopathol Soc Jpn. 1994;60:175–85.

Katsantonis D, Kadoglidou K, Dramalis C, Puigdollers P. Rice blast forecasting models and their practical value: a review. Phytopathol Mediterr. 2017;56(2):187–216.

Kumar J, Nelson RJ, Zeigler RS. Population structure and dynamics of *Magnaporthe grisea* in the Indian Himalayas. Genetics. 1999;152:971–84.

Kumar MP, Gowda DS, Moudgal R, Kumar NK, Gowda KP, Vishwanath K. Impact of fungicides on rice production in India. Fungicides-show cases of integrated plant disease management from around the world. London: IntechOpen; 2013.

Landschoot PJ, Jackson N. *Magnaporthe poae* spnov. a hyphopodiate fungus with a Phialophora anamorph from grass roots in the USA. Mycol Res. 1989;93:59–62.

Le MT, Arie T, Teraoka T. Population dynamics and pathogenic races of rice blast fungus, *Magnaporthe oryzae* in the Mekong Delta in Vietnam. J Gen Plant Pathol. 2010;76:177–82.

Li C, Li J, Sheng R, Fujita Y. Cross-fertility of rice blast fungus *Pyricularia oryzae*. Xi Nan Nong Ye Xue Bao. 1992;5:53–8.

Li C, Chen Q, Chen Z, Luo C, Hayashi N, Iso K. Distribution of the matting type in *Magnaporthe grisea* population pathogenic to rice in Yunnan Province. Sci Agric Sin. 1996;29:60–4.

Lowry RJ, Durkee TL, Sussman AS. Ultra structural studies of microconidium formation in *Neurospora crassa*. J Bacteriol. 1967;94:1757–63.

Maheshwari R. Microconidia of *Neurospora crassa*. Fungal Genet Biol. 1999;26:1–18.

Marcel S, Sawers R, Oakeley E, Angliker H, Paszkowski U. Tissue adapted invasion strategies of the rice blast fungus *Magnaporthe oryzae*. Plant Cell. 2010;22(9):3177–87.

Marciel JLN, Ceresini PC, Castroagudin VL, Zala M, Kema GH, McDonald BA. Population structure and pathotype diversity of the wheat blast pathogen, *Magnaporthe oryzae* 25 years after its emergence in Brazil. Phytopathology. 2014;104:95–107.

Mekwatanakarn P, Kositratana W, Phromraksa T, Zeigler RS. Sexually fertile *Magnaporthe grisea* rice pathogens in Thailand. Plant Dis. 1999;83:939–43.

Nalley L, Tsiboe F, Durand-Morat A, Shew A, Thoma G. Economic and environmental impact of rice blast pathogen (*Magnaporthe oryzae*) alleviation in the United States. PLoS One. 2016;11(12):e0167295.

Rathour R, Singh BM, Sharma TR. Sexual fertility in *Magnaporthe grisea* isolates from rice, finger millet and crabgrass in Himachal Pradesh. J Mycol Plant Pathol. 2004;34:230–2.

Ribot C, Hirsch J, Balzergue S, Tharreau D, Nottéghem JL, Lebrun MH, Morel JB. Susceptibility of rice to the blast fungus, *Magnaporthe grisea*. J Plant Physiol. 2008;165(1):114–24.

Saleh D, Xu P, Shen Y, Li C, Adreit H, Milazzo J, Ravigne V, Bazin E, Notteghem JL, Fournier E, Tharreau D. Sex at the origin: an Asian population of the rice blast fungus *Magnaporthe oryzae* reproduces sexually. Mol Ecol. 2012;21:1330–44.

Saleh D, Milazzo J, Adreit H, Fournier E, Tharreau D. South-East Asia is the center of origin, diversity and dispersion of rice blast fungus, *Magnaporthe oryzae*. New Phytol. 2014;201:1440–56.

Sesma A, Osbourn AE. The rice leaf blast pathogen undergoes developmental processes typical of root-infecting fungi. Nature. 2004;431(7008):582–6.

Singh RS. Plant diseases. Med-Tech International Publishers; 2017. p. 485–6.

Skamnioti P, Gurr SJ. Against the grain: safeguarding rice from blast disease. Trends Biotechnol. 2009;27:141–50.

Takan JP, Chipili J, Muthumeenakshi S, Talbot NJ, Manyasa EO, Bandyopadhyay R, Sere Y, Nutsugah SK, Talhinhas P, Hossain M, Brown AE, Sreenivasaprasad S. *Magnaporthe oryzae* populations adapted to finger millet and rice exhibit distinctive patterns of genetic diversity, sexuality and host interaction. Mol Biotechnol. 2012;50:145–58.

Tredway LP, Stevenson KL, Burpee LL. Mating type distribution and fertility status in *Magnaporthe grisea* populations from turf grasses in Georgia. Plant Dis. 2003;87:435–41.

Tucker SL, Talbot NJ. Surface attachment and pre-penetration stage development by plant pathogenic fungi. Annu Rev Phytopathol. 2001;39:385–417.

Ueyama A, Tsuda M. Formation of the perfect state in culture of *Pyricularia* sp. from some gramineaceous plants (preliminary report). Trans Mycol Soc Jpn. 1975;16:420–422.

Yaegashi H, Nishihara N. Production of the perfect stage in *Pyricularia* from cereals and grasses. Jpn J Phytopathol. 1976;42(4):511–5.

Yaegashi H, Udagawa S. The taxonomical identity of the perfect state of *Pyricularia grisea* and its allies. Can J Bot. 1978;56(2):180–3.

Zeigler RS. Recombination in *Magnaporthe grisea*. Annu Rev Phytopathol. 1998;36:249–75.

Zeigler RS, Scott RP, Leung SA, Teng PS. Proc. Symp. On Rice Blast Disease pp. 626, Univ. of Wisconsin-Madison, Madison, Wisconsin, USA, August 1994. CAB International, UK, and Inter. Rice Res. Inst., Los Banos, the Philippines. 1994.

Zhang H, Wu Z, Wang C, Li Y, Xu JR. Germination and infectivity of microconidia in the rice blast fungus *Magnaporthe oryzae*. Nat Commun. 2014;5:1–9.

Zickler D, Arnaise S, Coppin E, Debuchy R, Picard M. Altered mating-type identity in the fungus *Podospora anserina* leads to selfish nuclei, uniparental progeny, and haploid meiosis. Genetics. 1995;140:493–503.

Chapter 11
Understanding Pearl Millet Blast Caused by *Magnaporthe grisea* and Strategies for Its Management

Shweta Singh, Rajan Sharma, S. Chandra Nayaka, C. Tara Satyavathi, and Chandramani Raj

Contents

S. Singh · C. Raj
ICAR Research Complex for NEH Region, Sikkim Center, Tadong, Gangtok, India

R. Sharma (✉)
International Crops Research Institute for the Semi-Arid Tropics (ICRISAT), Patancheru, Hyderabad, Telangana, India
e-mail: r.sharma@cgiar.org

S. Chandra Nayaka
Pearl Millet Pathology Laboratory, DOS in Biotechnology, University of Mysore, Manasagangotri, Mysore, India

C. Tara Satyavathi
ICAR-AICRP on Pearl Millet, Jodhpur, Rajasthan, India

S. C. Nayaka et al. (eds.), *Blast Disease of Cereal Crops*, Fungal Biology,
https://doi.org/10.1007/978-3-030-60585-8_11

11.1 Introduction

Pearl millet {*Pennisetum glaucum* (L.) R. Br., [syn. *Cenchrus americanus* (L.) Morrone]} is a C_4 cereal crop grown in arid and semiarid regions of Asia and Africa in an area of 26 million hectares (Rai et al. 2009). It grows under harsh environmental conditions on infertile soils of low water-holding capacity, where other cereal crops are prone to fail (Manning et al. 2011). India is the largest producer of pearl millet in the world, with Rajasthan being the largest producer in India. The crop encounters a number of diseases which affect the crop adversely during the growth stages. The blast or leaf spot disease is caused by the fungus *Magnaporthe grisea* (Herbert) Barr [anamorph: *Pyricularia grisea* (Cooke) Sacc.]. It has emerged as one of the most destructive diseases of pearl millet in the recent past probably due to commercialization of new hybrids and also changing climate. The disease was first recorded in 1933 (Emechebe 1975) but very little was known about the pathogen biology, epidemiology, and diversity of pathogen populations. The disease has geographic distribution in India, African countries, the United States, and Singapore (on Napier grass) (Buckley and Allen 1951). In India, the disease was first reported in 1942 from Kanpur, Uttar Pradesh (Mehta et al. 1953). Once considered as a minor disease, incidence of blast has increased at an alarming rate in the recent past, particularly on commercial hybrids affecting both forage and grain production of pearl millet (Sharma et al. 2013). The disease causes losses of both grain yield (Timper et al. 2002) and forage yield (Wilson and Gates 1993). Due to widespread occurrence and increasing threat to pearl millet production, blast has now been identified as top priority of pearl millet improvement programs both in public and private sectors.

11.2 Distribution

Pearl millet blast is the second most devastating disease after downy mildew affecting aerial parts of the plant at all stages of its growth starting from the seedling stage (causing lesions and premature drying of young leaves) to the panicle resulting in neck and grain blast. It is known to cause economic losses in major pearl millet-growing regions of the world. In India, the worst affected states are Rajasthan, Gujarat, Maharashtra, Uttar Pradesh, and Haryana, though the disease can be observed in all the pearl millet-producing areas in India; leaf blast is considered as a serious disease in southern coastal plains of the United States where infection from this disease has been found to have significant adverse effects on green forage yield and digestible dry matter (Wilson and Gates 1993). It is also common and severe in most of the African countries wherever pearl millet is grown (Wilson et al. 1989; Werder and Manzo 1992; Marley et al. 2002; Lubadde et al. 2014).

11.3 Current Status of Disease in India

The incidence of blast disease in India has been observed since 1970, and an increase has been observed in most of the pearl millet-growing regions of the country, namely Rajasthan, Maharashtra, Gujarat, Uttar Pradesh, Haryana, and Madhya Pradesh. The disease has become widespread during the past decade, and is being observed on most of the prominent cultivars grown in the major pearl millet-producing areas with high disease severity under warm and humid conditions (AICMIP Annual Reports, 2002–2019; http://www.aicpmip.res.in/) (Table 11.1).

11.4 The Pathogen

Magnaporthe grisea was placed in the taxonomic class Ascomycota or existed as related asexual forms (Agrios 1997). *M. grisea* was further included in the taxonomic class Pyrenomycetes because it produces asexual spores in a flask-shaped structure called a perithecium. The systematic position of the fungus is given below:

Domain: Eukaryota
Kingdom: Fungi
Phylum: Ascomycota
Subphylum: Pezizomycotina
Class: Sordariomycetes
Subclass: Sordariomycetidae
Order: Magnaporthales Family: Magnaporthaceae
Genus: *Magnaporthe*
Species: *Magnaporthe oryzae/grisea*

The pathogen has been described as genus *Pyricularia* (Cooke) Sacc. (anamorphic: Magnaporthaceae) and it was established by Saccardo (1880) with the type species, *P. grisea* which was originally described from crabgrass (*Digitaria sanguinalis* L.). The name "Pyricularia" refers to the pyriform shape of the conidia. Cavara (1892) subsequently described *P. oryzae* Cav. from rice (*Oryza sativa* L.), a taxon with similar morphology to *P. grisea*. Though two taxons are morphologically similar they have been maintained as separate species. It was Rossman et al. (1990) who advocated to synonymize *P. grisea* with *P. oryzae* and therefore grouped these two anamorphs under the teleomorph *Magnaporthe grisea* (Hebert) Barr. However, *Magnaporthe oryzae* has now been segregated as a distinct species from *M. grisea* based on a multilocus phylogenetic analysis and on mating properties of the strains (Klaubauf et al. 2014). *M. grisea* isolates are pathogenic on *Digitaria* and related grasses, and *M. oryzae* is associated with rice and diverse grasses of agricultural significance (Couch and Kohn 2002).

Table 11.1 Blast disease incidence on pearl millet observed in on-farm surveys conducted during 2002–2019

Year	Range of blast incidence (%)	Indian states
2002–2003	0.5–60	Gujarat, Rajasthan
2003–2004	2–80	Gujarat
2004–2005	2.5–65	Gujarat, Madhya Pradesh
2005–2006	1–69	Gujarat, Madhya Pradesh, Maharashtra
2006–2007	2–20	Gujarat, Madhya Pradesh, Karnataka
2007–2008	2–15	Gujarat, Madhya Pradesh, Karnataka
2008–2009	1.3–50	Gujarat, Madhya Pradesh, Karnataka
2009–2010	1.0–20	Rajasthan, Gujarat, Madhya Pradesh, Maharashtra, Karnataka
2010–2011	2.0–80	Rajasthan, Madhya Pradesh, Maharashtra, Tamil Nadu
2011–2012	1.0–80	Rajasthan, Gujarat, Madhya Pradesh, Maharashtra, Tamil Nadu
2012–2013	1.0–90	Rajasthan, Gujarat, Madhya Pradesh, Maharashtra, Tamil Nadu
2013–2014	2.0–70	Rajasthan, Gujarat, Madhya Pradesh, Maharashtra, Tamil Nadu, Karnataka
2014–2015	1.0–50	Rajasthan, Gujarat, Madhya Pradesh, Maharashtra, Tamil Nadu, Karnataka
2015–2016	1.0–60	Rajasthan, Gujarat, Madhya Pradesh, Maharashtra, Tamil Nadu, Karnataka
2016–2017	1–90	Karnataka, Maharashtra, Madhya Pradesh
2017–2018	5–80	Maharashtra, Karnataka, Madhya Pradesh
2018–2019	5–90	Rajasthan, Maharashtra, Madhya Pradesh, Tamil Nadu
2019–2020	5–90	Karnataka, Rajasthan, Maharashtra, Madhya Pradesh

Source: AICMIP Annual Reports, 2002–2019

The asexual stage described by the name *Pyricularia oryzae* (formerly called *P. grisea*) is the most common spore form of the fungus. These spores, called conidia, are produced abundantly on lesions and in culture on specialized stalks, called conidiophores. Conidiophores are simple, septate, basal portion being relatively darker. Conidia are pyriform in shape and hyaline in color produced acrogenously, one after another. Conidium is three celled, with the middle cell being much wider and darker and end cell germinating, giving out germ tubes. The conidia usually measure 17.5–30.8 × 5.9–8.8μm (Mehta et al. 1953). Hyphae are hyaline and septate; however, as the fungus gets older, the hypha becomes brown. Formation of intercalary or terminal chlamydospores is common, which are globose, thick walled, and olive brown.

The teleomorphic phase *Magnaporthe grisea* is an ascomycete fungus producing flask-shaped perithecial bodies that carry unitunicate asci bags containing ascospores, the products of meiosis in abundance. Asci can be dissected to liberate the ascospores, which are arranged as unordered octads (four pairs of spores representing the products of meiosis that have undergone a subsequent mitotic division) or as larger populations of randomly selected ascospores which are four celled and spindle shaped (Barr 1977; Hebert 1971). This fungus is considered to be heterothallic with a bipolar mating system (mating controlled by two different alleles at a single locus) with additional genes controlling the sexual cycle (Talbot 2003; TeBeest et al. 2007). The occurrence of sexual stages of rice blast fungus has been reported in Kerala and central Himalaya in India (Kumar et al. 1999; Brindha et al. 1999). However, there is no report so far on the occurrence of sexual stage of the fungus from pearl millet.

As a hemi-biotrophic organism, *M. grisea* initially develops an intimate relationship with its host in compatible interactions, and presumably it is able to deal with any defenses the host may mount. Subsequently, host cells die resulting in characteristic "blast" disease symptoms.

11.5 Host Range

The members of *Magnaporthe grisea* (Hebert) Barr complex are heterothallic, filamentous fungus pathogenic to almost 40 plant species in 30 genera (137 members) of Poaceae including rice, wheat, barley, pearl millet, finger millet, foxtail millet, and several grasses (Ou 1980, 1985; Murakami et al. 2000; Inukai et al. 2006). The pathogen is highly variable, but highly specialized in its host range. The pathogen also survives on other graminaceous hosts such as *Agrostis palustris*, *Brachiaria mutica*, *Cyperus rotundus*, *Eleusine indica*, *Eragrostis* sp., and *Panicum miliaceum* (Lanoiselet and Cother 2005). Although these hosts support pathogenic forms of the fungus that are normally restricted to one or a small number of very closely related species, the organism has been best studied as the causal agent of rice blast disease (Mackill and Bonman 1986). Thus, *P. grisea* strains from rice or any other hosts generally do not cross infect each other in nature (Thakur et al. 2011).

The *Pennisetum* genus is quite diverse with about 100 species. However, there is no scientific study to prove the susceptibility of all these genera to *M. grisea* infection. The scientific evidences report that *Pennisetum glaucum*, *P. squamulatum*, *P. macroforum*, *P. pedicellatum* (Saikia et al. 1982), *P. ciliare* (Perrott and Chakraborty 1999), *P. purpureum* (Buckley and Allen 1951), and *P. violaceum* (Sharma et al. 2020) are susceptible to infection by *M. grisea*.

11.6 Symptoms

The blast disease of pearl millet is often referred to as gray spot of leaf and stem. The disease appears as grayish, water-soaked foliar lesions that enlarge and become necrotic, resulting in extensive chlorosis and premature drying of young leaves. The lesions usually start near the leaf tips or leaf margins or both and extend down toward the outer edges (Wilson et al. 1989). The lesions are usually confined to inter-veinal spaces on the foliage. Lesions grow and coalesce to cover large surface areas and cause necrosis of tissues. Depending on the resistance level of host cultivar and environmental conditions, the lesion size varies from small, roundish, elliptical, diamond-shaped to elongated lesions. The center of the lesion is gray and water soaked while fresh but turns brown surrounded by a chlorotic halo, which turns necrotic, giving the appearance of concentric rings (Kato 2001). These symptoms appear from seedling to flowering stage on leaf, stem, and boot leaf. In case of susceptible cultivars the entire foliage gives a burnt appearance. Severely infected plants produce few shriveled grains in the blasted florets. This disease becomes more severe during humid weather conditions especially with dense plant stands (Thakur et al. 2011).

11.7 Epidemiology

Weather variables, particularly relative humidity, leaf wetness duration, and temperature, play a major role in influencing infection and disease development in any host-pathogen system. Blast disease has the potential to cause severe crop losses in pearl millet when environmental conditions are favorable for disease development. Therefore, information on relationship between weather variables and blast disease could be used to refine techniques to screen germplasm/breeding lines for resistance. Based on limited information on weather variables conducive for blast development, both greenhouse and field screening techniques for blast resistance have been developed (Thakur et al. 2011; Sharma et al. 2013). Improved knowledge of the effect of interaction of host cultivar with weather variables, pathogenic strains, and crop growth stages would be helpful in understanding and predicting the disease epidemics. These factors are more relevant with polycyclic, airborne pathogens like *M. grisea*.

Magnaporthe grisea is a hemi-biotrophic pathogen which primarily grows biotrophically and later transforms into necrotrophs, thus killing the infected tissues (Perfect and Green 2001; Munch et al. 2008). *M. grisea* itself maintains both biotrophic and necrotrophic stages of the pathogen and utilizes it for invading the foliar tissues of the host (Kankanala et al. 2007).

Though information available on the epidemiology of pearl millet blast is scanty, rice blast is a widely studied disease. Thus, information available on the epidemiology of rice blast might help in understanding the epidemiology of pearl millet blast. In general, long periods of leaf wetness, high relative humidity (>90%), high temperatures of 25–28 °C, cloudy sky, frequent rain and drizzle, presence of conidial spores of blast in atmosphere/air, and high nitrogen fertilizer application favor the blast disease development in rice (Lamey 1970; Kim and Kim 1993; Teng 1994; Ou 1985; Kato 2001).

The pyriform conidia on seed and the mycelial strands on the surface of infected seeds, diseased straw, and stubbles serve as primary source of inoculum and infect young seedlings. The infected leaves become reservoir of inoculum in 8 days of pathogen establishment on young leaf tissues, and sporulate liberating large number of airborne conidia in the air (Chandra Nayak et al. 2017). The airborne conidia act as secondary source of inoculum which falls on the surface of seedlings or the plants from tillering to flowering stage that infect a large number of plants. Under dry conditions (at room temperature) conidia can survive for a year and mycelium for almost 3 years (Shetty et al. 2009; Kato 2001).

A single leaf lesion carries multiple conidiophores which can release up to 20,000 conidia for 20 days during night. The foliar infection is initiated by attachment of a three-celled conidium of *M. grisea* to the pearl millet leaf cuticle. In the presence of moisture, the conidium germinates on the leaf surface, and produces a germ tube which develops a melanized appressorium which facilitates the penetration peg to mechanically pierce the cuticle on the host cell surface (Wilson and Talbot 2009). On the host surface, the fungus may also respond to cutin monomers, as cis-9, 10-epoxy-18-hydroxyoctadecanoic acid, or lipid monomers, such as 1,16-hexadecanediol, that are powerful inducers of appressorium development (Talbot 2003; Ebbole 2007). Inside the cell lumen, invasive hypha (IH) which is bulbous and intensively dividing is formed and gets bordered by a plant-derived membrane that bifurcates the IH and host cytoplasm, and causes a lesion, a distinguishing feature of biotrophy. The biotrophic nature of early blast infection is also suggested by recent evidence that movement of the fungus from cell to cell may occur by means of plasmodesmata; the fungus appears to seek out pit field sites when invasive hyphae move to adjacent epidermal cells. Furthermore, the fungus appears to be excluded from entering stomatal guard cells, which lack plasmodesmata, consistent with this mode of cell-to-cell spread (Kankanala et al. 2007). This stage is followed by the lesions becoming necrotic and starting coalescing which is the necrotrophic stage of the pathogen. Epidemic of blast disease is caused by the large number of conidia formed from the disease lesions. The disease is polycyclic with a spore-to-spore cycling time of about 7 days (Skamnioti and Gurr 2009).

The sexual or teleomorphic phase of the fungus initiates in the presence of two opposite mating types or hermaphroditic isolates when they come into contact. As an ascomycete, it produces hyaline, three-septate, fusiform-shaped, spindle-shaped ascospores, and pigmented cells in the center (Cannon 1994). The ascospores are produced in asci which are unitunicate in *Magnaporthe*. This fungus is considered to be heterothallic with a bipolar mating system (mating controlled by two different alleles at a single locus) with additional genes controlling the sexual cycle (TeBeest et al. 2007). The mating contact results in the formation of bulbous structure with an elongated neck called as perithecium. Inside the perithecium specialized spore-forming structures called asci are developed where several ascospores are formed. The ascospores germinate and form the hyphal structures which cause lesions on the host tissues (Chandra Nayak et al. 2017).

11.8 Pathogen Characterization

11.8.1 Cultural and Morphological Characterization

The fungus *M. grisea* produces three-celled, pyriform macroconidia in the imperfect stage and four-celled, spindle-shaped ascospores in the perfect stage (Barr 1977; Hebert 1971; Kato and Yamaguchi 1982). The fungal mycelium in cultures is aerial or submerged, hyaline or olivaceous, 1.5–6.0μm in width, with one to many conidiophores, fasciculate, simple, or rarely branched, 2–4 septate, not or slightly constricted at septa; at first monosporic, then pleurogenous on sympodium, olivaceous to fuliginous, base swollen, dark colored and becoming lighter color toward the apex. Asexual conidia are variable in size and shape, terminal, pyriform to obclavate, mostly three celled with a small appendage on the rounded base cell, apex narrowed: two septate, rarely 1–3 septate, not or slightly constricted at septa, almost hyaline to pale olive, approximately 17.5–30.8 × 5.9–8.8μm (Mehta et al. 1953). The conidiophores of the fungus are produced in clusters from each stroma. They are rarely solitary with 2–4 septa. The basal area of the conidiophores is swollen and tapers toward the apex. The conidia germinate from apical or basal cell and less frequently from middle cell and measure up to 20–22 × 10–12μm. The conidia are translucent and slightly darkened, obclavate and tapering at the apex with or without constriction at septa, branched, and 3–5μm in width (Shirai 1896; Sawada 1917; Nishikado 1926). In 1994 Kato et al. (1994) found microconidia in *M. oryzae* cultures on artificial media. Morphologically distinct from macroconidia, they were characterized by unique features such as a single cell with no septum; small size, 5–8 (mean 6) μm long and 0.5–0.8 (0.7) μm wide; with one nucleus; lunate; and hyaline. Kato et al. (1994) demonstrated that the small conidia were a third type of spores produced by *Magnaporthe oryzae* during its life cycle and named them microconidia.

Hebert (1971) was the first to publish the details of morphology of the *Magnaporthe grisea* species, and Yaegashi and Hebert (1976) reported on perithecial development but did not obtain fertile crosses from isolates of *P. oryzae.* However, Ueyama and Tsuda (1975) reported formation of a perfect state resembling *Ceratosphaeria grisea* (the basyonym) from crosses of *P. oryzae* and a species of *Pyricularia* isolated from *Eleusine*. The teleomorph of the fungus has been produced on artificial media through crosses between *P. grisea* × *P. grisea*, *P. oryzae* × *P. grisea*, and *P. oryzae* × *P. oryzae* (Hebert 1971, 1975; Kato and Yamaguchi 1982). The existence of hermaphroditic isolates in some parts of the world denied the absence of sexual reproduction under field conditions in this fungus.

Silue and Notteghem (1990) attempted to produce teleomorph of *P. oryzae* on rice plants in growth chambers. They inoculated rice plants with conidial suspension of compatible strains of two complementary mating types which crosses produced greater number of perithecia on rice-inoculated plants (8–15 days) in a growth chamber at 20 °C under fluorescent lights (Yaegashi and Yamada 1986). The formation of perithecia could be observed on the dead tissues of rice stems on 15th day in growth chamber. However, production of perithecium by the pathogen under laboratory conditions has not been observed for pearl millet blast system.

11.8.2 Molecular Characterization

The evolutionary potential of *M. grisea* enables the pathogen to develop new races/strains after the continuous selection pressure from resistant host cultivar that eventually challenges the breeding for resistance in pearl millet. Therefore, successful disease management strategies essentially need knowledge of genetic structure and dynamics of pathogen population to prevent the breakdown of resistance genes (Lavanya and Gnanamanickam 2000). However, scanty literatures are available on molecular diversity of *M. grisea* populations adapted to pearl millet. In general, the most commonly used DNA-based markers for genetic diversity studies include randomly amplified polymorphic DNA (RAPD), amplified fragment length polymorphism (AFLP), sequence characterized amplified region (SCAR), and microsatellites or SSR markers. However, Kiran Babu et al. (2013) reported little or no variation in *M. grisea* populations infecting pearl millet. The study on molecular diversity of *M. grisea* isolates associated with pearl millet with nine URP markers (URP-1F, 2F, 4R, 6R, 9F, 17R, 25F, 30F, and 38F) revealed average gene diversity of 0.21 ± 0.10 over loci (Yella Goud 2015). The results showed highest polymorphic information content (PIC) value of 0.50 with three primers URP4R, URP9F, and URP25F and PIC range varied from 0.32 to 0.50 for each marker with an average of 0.43. Analysis of molecular variance (AMOVA) indicated significant genetic variation between and within the 65 isolates of *M. grisea*. The genetic variation of 84% was observed due to difference in the isolates whereas 16% variation was noticed ($FST = 0.1594$) due to differences between populations of *M. grisea* adapted to pearl millet and non-pearl millet hosts.

11.8.3 Pathogenic Variability

The pathogenic variability in *M. grisea* populations infecting rice, finger millet, foxtail millet, wheat, and other weed grasses has been widely studied and reported (Prabhu et al. 1992; Nakayama et al. 2005; Takan et al. 2012). The presence of several pathogenic races of *M. oryzae* adapted to rice has been confirmed and this pathogenic diversity in the pathogen causes frequent breakdown of blast resistance in rice varieties (Suh et al. 2009). The frequent change in the pathogenic races of *M. grisea/oryzae* can be attributed to various mechanisms like sexual recombination, heterokaryosis, parasexual recombination, and aneuploidy (Kang and Lee 2000). The first study on pathogenic variability in the pearl millet-infecting populations of *M. grisea* in India has shown the presence of five different pathotypes based on the reaction of 25 isolates on ten differential lines of pearl millet. Among five different pathotypes represented by five isolates, isolate Pg118 from Rewari, Haryana, was the most virulent followed by Pg056 from Gotan, Rajasthan (Sharma et al. 2013). This differential set has been further refined by adding and removing some lines, and used to discern the variability in about 80 isolates collected from different pearl millet-growing areas in India (Sharma et al. 2019). The management of disease through host plant resistance necessitates identification of diverse pathotypes of the pathogen, and sources of resistance against these pathotypes. The pathogenic variability studies of *M. grisea* have led to the identification of diverse pathotype isolates which are currently being used in the greenhouse screening at ICRISAT, Patancheru, India, for the development of blast-resistant hybrid parental lines, and hybrids of pearl millet (Sharma et al. 2019).

11.9 Mating Types

M. grisea is a heterothallic fungus and the pathogen populations comprise two distinct mating types (MAT1-1 and MAT1-2), and sexual reproduction is possible only between the two opposite mating types (Couch and Kohn 2002; Talbot 2003). It is a highly variable fungus with high degree of pathogenic variation (pathotypes) in the field (Silva et al. 2009; Kumar et al. 1999). Assessment of mating-type alleles has been used as a marker to measure population diversity in this pathogen (Zarrinnia et al. 2012; Silva et al. 2009; Kumar et al. 1999; Kang et al. 1994; Urak et al. 2008). The mating system in *M. grisea* is controlled by a single master locus MAT 1 (Coppin et al. 1997; Yoder et al. 1986) where both MAT1-1 and MAT1-2 are idiomorphs required for sexual recombination. These two mating types are present in natural populations of rice-growing areas; however no teleomorphs were reported in nature (Kato 1974; Yaegashi 1977). MAT1-1 and MAT1-2 have been reported from rice, foxtail millet, finger millet, and barley from different parts of the globe (Kang et al. 1994; Urak et al. 2008; Kotasthane et al. 2004; Zeng et al. 2009). However, one mating type usually dominates in natural *M. oryzae* populations (Cai et al. 2005; Consolo et al. 2005; Nottéghem and Silue 1992; Zeigler 1998) and this highly skewed mating type ratio is believed to be a major block to sexual recombination under field conditions (Consolo et al. 2005; Mekwatanakarn et al. 1999; Sommerhalder et al. 2006). Even where both mating type isolates coexist, other barriers, in particular the sexual and fertility status of the fungus, can also hinder successful mating (Leslie and Klein 1996; Sommerhalder et al. 2006; Tredway et al. 2003). The distribution of mating types has been used to characterize the genetic structure and dynamics of heterothallic populations (Consolo et al. 2005; Mekwatanakarn et al. 1999; Nottéghem and Silue 1992; Sommerhalder et al. 2006; Tredway et al. 2003; Zeigler 1998). MAT1-1 is usually prominent in rice populations and presence of high frequency of MAT1-2 seems to be a dominant feature of non-rice isolates as supported by the reports of several workers (Kotasthane et al. 2004).

The mating types in conventional approach were determined based on the appearance of matured perithecia in a cross between known tester and unknown strain on culture media which demanded technical expertise and was time consuming (Kumar et al. 1999; Priyadarshini et al. 1999; Nottéghem and Silue 1992). Development of PCR-based amplification techniques using *Mat* gene-specific primers paved way for a rapid method to explore the mating-type population of *M. oryzae/grisea* (Zheng et al. 2008; Bao-Hua et al. 2005; Dong-mei et al. 2005). The frequencies of MAT1-1 and MAT1-2 idiomorphs in different hosts have been easily assessed by amplicons produced after PCR using specific primers for MAT1-1 and MAT1-2 (Takan et al. 2012; Xu and Hamer 1995; Dong-mei et al. 2005; Urak et al. 2008). The PCR-based approach permits convenient mating-type assessment that is independent of other incompatibility factors. Using genomic subtraction approach, Kang et al. (1994) cloned and sequenced 2.5 kb MAT I and 3.5 kb MAT 2 idiomorphs from *M. grisea* and reported fertility characteristics of strains containing the opposite mating-type idiomorph, and of dual-mater strains that contain both mating-type idiomorphs. The identification of mating types of field isolates in the future can be facilitated by using molecular probes instead of crossing to tester strains (Kang et al. 1994).

11.10 Genome Organization

The draft genome of the pearl millet *M. grisea* consists of 9649 coding DNA sequences (CDSs) with a total genome size of 49.90 Mb including 341 scaffolds. Among these, a total of 849 CDSs were annotated to be involved in pathogenicity, virulence, and effector genes (Prakash et al. 2019). However, the genome of rice blast *M. grisea* has elucidated in detail the mechanism of pathogenicity in this widely studied pathogen. The genome sequence information indicates that the pathogen is capable of secondary metabolite production. The three mitogen-activated protein kinases (MAPK) regulate appressorium development, penetration peg formation, and adaptation to hyperosmotic stress. Among the three, two MAPK pathways are involved in controlling virulence-associated development of *M. grisea*. The appressorium formation in *M. grisea* is regulated by the core of Pmk1 MAPK (pathogenicity MAPK). A few components of Pmk1 pathway such as MAPK are involved in mating and pathogenesis, and the Mst1 and Ste12-related transcription factors are dispensable for mating altogether (Xu 2000). Analysis of the *M. grisea* genome also suggests that germinating spores possess considerable versatility in their capacity to synthesize glycerol up to 3M in the appressorium and generate enough turgor pressure to break plant cuticle (Thieringer and Kunau 1991).

There are 23 genes predicted to encode polyketide synthases (PKS), and 3 of these genes in particular are upregulated during infection-related development. Their non-ribosomal peptide synthetases (NRPS) are also very well represented in the genome. They catalyze production of cyclic peptides including numerous toxins. There exist two distinct subclasses: the one which exists separately and the others that are fused to a PKS (PKS-NRPS). Therefore, in the *M. grisea* genome there are six likely NRPS genes and eight PKS-NRPS genes of which six are full-length PKS-NRPS genes and two PKS-NRPS genes are with a truncated NRPS domain. The large number, and expression profile, of PKS and NRPS genes in *M. grisea* is consistent with the requirements of a fungal pathogen in adapting to diverse environments, perturbing host metabolism and ultimately causing plant cell death (Bohnert et al. 2004). The pathogen *M. grisea* is predicted to secrete 739 proteins which were identified as secreted proteomes of the fungus that are crucial in perceiving and responding to the environmental cues. A number of these proteome families are predicted to encode enzymes for degradation of the plant cell wall and cuticle; eight genes in *M. grisea* putatively encode cutinases-methyl esterases that degrade cutin, the waxy polymer that forms the leaf cuticle (Dean et al. 2005). During infection process, such genes get significantly upregulated. An investigation into the *M. grisea* genome revealed three families of putatively secreted, cysteine-rich polypeptides which are putative effector proteins and have been recognized as

PAMPs (Joosten et al. 1994). The genome of the sequenced *M. grisea* strain 70–15 contains four known *M. grisea* avirulence genes: *AVR-Pita*, *ACE1*, *PWL2*, and *PWL3*. However, no orthologues of *M. grisea PWL1*, *PWL4*, or *AVR1-CO39* or well-characterized AVR genes from other pathogenic fungi including *Avr2*, *Avr4*, *Avr9*, *ECP2*, *ECP3*, and *ECP5* were found which highlights the diversity in the fungal avirulence and lack of conservation gene (Steiner-Lange et al. 2003).

11.11 Mechanism of Pathogenicity

Magnaporthe complex is a typical phytopathogen which produces a specialized infection cell called an appressorium to infect its host (Oh et al. 2008). The process of *M. grisea* infection starts when a conidium lands on the pearl millet leaf surface. After attachment and germination of spore on the host surface, the emerging germ tube perceives physical cues, such as surface hardness and hydrophobicity, as well as chemical signals that trigger appressorium formation (Lee and Dean 1993a; Flaishman and Kolattukudy 1994; Gilbert et al. 1996). The genome organization studies on *M. grisea* have revealed upregulation of two specific CFEM-GPCR (conserved fungal specific extracellular membrane-G-protein-coupled receptors) during appressorium formation, one of which is pth11 also required for pathogenesis (DeZwaan et al. 1999; Talbot 2003; Dean et al. 2005). The contents of spore are mobilized into the appressorium which upon maturity becomes firmly attached to the plant surface and a dense layer of melanin is laid down in the wall except across a pore at plant surface. The turgor pressure inside the appressorium increases up to 8 MPa (de Jong et al. 1997) which is maintained by the melanin deposition which enables the penetration hyphae to penetrate at the pore, which is driven through the plant cuticle into the underlying epidermal cells (Staples et al. 1976; Bourett and Howard 1990; Howard and Valent 1996; de Jong et al. 1997; Shaw et al. 1998; Thines et al. 2000). In an incompatible interaction, resistance gene products recognize corresponding avirulence gene products from the invading pathogen and initiate an array of defense responses to restrict pathogen growth (Ahn et al. 2005). In case of compatible interaction, the host plant mobilizes defense responses much later, resulting in visible coalescing lesions on the leaf surface. The conidiophores emerge from these lesions producing numerous new conidia that are able to start a second round of infection. The gene products of the pathogen expressed during infection process may serve as pathogenicity factors required for evasion of the host's defense and successful colonization.

Highly conserved signaling networks that transfer cues from the environment to the nucleus play a crucial role in regulating pathogen-host interactions. The pathways that have been found to be essential for appressorium formation and function in the fungus are mitogen-activated protein kinase (MAPK), cyclic AMP (cAMP), and to some extent Ca^{2+} signaling (Lee and Dean 1993b; Xu and Hamer 1996; Choi and Dean 1997; Dean 1997; Adachi and Hamer 1998; Lee and Lee 1998). In addition, the cAMP signaling pathway regulates several other aspects of fungal growth and development, including nutrient sensing and cell morphogenesis (Borges-Walmsley and Walmsley 2000; D'Souza and Heitman 2001; Lee et al. 2003). The core pathways are found to be highly conserved and a little is known about the downstream genes and pathways that govern infection-related morphogenesis.

11.12 Management of Pearl Millet Blast

Management of pearl millet blast could be designed in integrated manner by keeping the following points in mind: (1) survival of pathogen as chlamydospores or as free saprophytic mycelium on external seed surface or in soil/leaf debris and (2) secondary spread of *M. grisea* conidia within and between pearl millet fields. The various management strategies that can be used to control pearl millet blast are given below:

11.12.1 Cultural Practices

The manipulations of environment in cultural practices are aimed toward providing maximum advantage to the host as compared to the pathogen. Planting of disease-free seeds or surface-sterilized seed, avoiding excess nitrogen-based fertilizers in the field, early sowing, field sanitation after harvesting the crop, and burning or composting of diseased straw and stubble are essential components of cultural methods for pearl millet blast management. The harvesting of seed must be done from disease-free areas or disease-controlled areas. However, these economical cultural measures are not effective under favorable conditions of blast disease.

11.12.2 Host Plant Resistance

The use of host plant resistance (HPR) is the most suitable approach in managing the pearl millet diseases as the crop is grown under resource-poor conditions by deprived farmers in semiarid tropics. Development of durable resistant varieties with broad genetic resistance is always considered as the most economical and eco-friendly method of plant disease management and success of such programs depends on the identification of stable resistance sources that can be utilized subsequently (Nagaraja et al. 2007). The screening of a large number of germplasms at several locations against the predominant strains/populations of blast pathogen reveals the stable resistance in germplasm. To ease the resource and time consumption in large-scale evaluation of the entire germplasm collection, Frankel and Brown (1984) proposed the concept of a core collection (10% of the entire collection) which represents over 70% of the genetic variation within the entire collection. However, to further reduce this number, Upadhyaya and Ortiz (2001) gave the concept of mini-core collection, comprising 10% of the core or 1% of the entire collection, which still represents most of the useful variation in a crop species. The 238 germplasm accessions of pearl millet mini-core were screened under greenhouse conditions against five *M. grisea* pathotype isolates (Pg118, Pg119, Pg56, Pg53, and Pg45). Resistance to multiple pathotypes (two or more) was recorded in several accessions, while three accessions (IP 7846, IP 11036, and IP 21187) exhibited resistance to four of the five isolates.

Sources of blast resistance were identified in the United States and efforts have been made to incorporate resistance into improved cultivars and elite breeding lines (Hanna et al. 1988; Wilson and Hanna 1992a, b). The elite B-lines and R-lines (863B, ICMB 01333, ICMB 01777, ICMB 02111, ICMB 03999, ICMB 93222, ICMB 97222, ICMR 06222, ICMR 06444, and ICMR 07555) developed at ICRISAT have been identified with high levels of resistance to blast (Thakur et al. 2009). In another study, Yella Goud et al. (2016) screened 160 designated B-lines of pearl millet for blast resistance against five pathotype isolates (Pg 45, Pg 53, Pg 56, Pg 118, and Pg 119) of *M. grisea* and found seedling-stage resistance in 23 lines of pearl millet to 3–5 pathotypes. Among 23 resistant B-lines, 9 (81B, ICMB 88004, ICMB 92444, ICMB 97222-P1, ICMB 02111, ICMB 06444, ICMB 07111, ICMB 09333, and ICMB 09999) showed resistance to all the five pathotype isolates. Crop wild relatives have also been tapped for the identification of novel sources of disease resistance. For identification of diverse sources of blast resistance, 305 accessions of *P. violaceum*, a wild relative of pearl millet, were screened under greenhouse conditions against five pathotype isolates (Pg 45, Pg 53, Pg 56, Pg 118, and Pg 119). Based on the mean blast score, 17 accessions (IP 21525, 21531, 21536, 21540, 21594, 21610, 21640, 21706, 21711, 21716, 21719, 21720, 21721, 21724, 21987, 21988, and 22160) were found resistant (score ≤ 3.0) to all the five pathotypes and 24 accessions were resistant to any four pathotypes of *M. grisea* (Sharma et al. 2020). Though resistance sources have been identified against blast disease, the resistant germplasm becomes susceptible over the period of time after evolution of new strains/races of the pathogen and thus search continues for the identification of new sources of high levels of resistance.

Information on inheritance of disease resistance facilitates the introgression of resistance genes in elite cultivars. It is relatively easy to transfer simply inherited traits. Several efforts have also been made to understand the inheritance of resistance to *M. grisea* and based on the reaction of F_1s it was concluded that resistance is governed by either dominant or partially dominant genes in pearl millet (Wilson et al. 1989; Wilson and Hanna 1992a, b; Gupta et al. 2012; Singh et al. 2018). The segregation pattern of resistance in F_2 and backcross generations indicated the governance of resistance by three independent dominant resistance genes (Hanna and Wells 1989; Wilson et al. 1989) in wild accession of pearl millet. Further, the study of inheritance pattern in pearl millet to the Indian isolate of *P. grisea* revealed the presence of single dominant gene governing resistance (Gupta et al. 2012; Singh et al. 2018). The resistance sources identified so far are being used in the pearl millet breeding program at ICRISAT to transfer blast resistance in elite hybrid parent lines.

11.12.3 Chemical Control

Although blast is primarily managed through host plant resistance, the pathogen develops new pathogenic races within few years that leads to breakdown of resistance in most of the commercial cultivars. In the absence of blast-resistant cultivars, the disease can be best managed with chemical fungicides (Sharma et al. 2018). Two field sprays of carbendazim @ 0.05% was recommended at the periodic interval of 15 days for controlling the blast diseases in pearl millet (Singh and Pavgi 1977; Lukose et al. 2011). However, this fungicide was not found effective in the study conducted by Sharma et al. (2018). The field experiment on the evaluation of newer fungicides by Joshi and Gohel (2015) showed that two foliar sprays of tricyclazole @ 0.05%, iprobenfos @ 0.1%, or isoprothiolane @ 0.05% at an interval of 15 days commencing from the first initiation of disease effectively manage the blast disease. Roopadevi and Patil (2017) evaluated 20 different fungicides under *in vitro* conditions against blast disease of pearl millet and found cent percent inhibition of pathogen growth with mancozeb 75% WP @ 0.1%, tricyclazole 75% WP @ 0.025%, carbendazim 12% + mancozeb 63%, and tebuconazole 50% + trifloxystrobin 25% @ 0.05%. The study of Sharma et al. (2018) on nine different fungicides (chlorothalonil, tricyclazole, hexaconazole, kasugamycin, benomyl, carbendazim, tebuconazole + trifloxystrobin, propiconazole, and metalaxyl + mancozeb) revealed that blast disease of pearl millet can be effectively managed with three foliar sprays of tebuconazole + trifloxystrobin @ 0.04% or propiconazole @ 0.1%. However, none of these fungicides was effective as seed treatment. Kaurav et al. (2018) conducted field experiment on the management of pearl millet blast with foliar application of fungicides (iprobenfos 48 EC @ 0.1%, tricyclazole @ 0.1%, azoxystrobin 25 EC @ 0.05%, propiconazole @ 0.05%, trifloxystrobin + tebuconazole @ 0.05%, and hexaconazole @ 0.1%) and observed the satisfactory level of disease protection by application of four sprays of trifloxystrobin + tebuconazole @ 0.05% after commencement of disease at fortnight intervals. Trifloxystrobin + tebuconazole has also been tested in the multilocation testing, and 2–3 sprays of this fungicide have been found effective in controlling pearl millet blast.

11.12.4 Biological Control

Biological control of a plant disease involves the use of living organism or naturally extracted or fermented products from various sources to inhibit the activity of plant pathogen. The bio-agents have been found to control the blast pathogen either through biocontrol or through induced systemic resistance (Taguchi et al. 2003). Several studies on the management of pearl millet blast have been conducted with *Trichoderma viride*, *Trichoderma harzianum*, *Pseudomonas fluorescens*, *Bacillus subtilis*, and *Bacillus pumilus*. Roopadevi and Patil (2017) evaluated *T. harzianum*, *P. fluorescens*, and *B. subtilis* either individually or in combination and found that the combination product of *T. harzianum* + *B. subtilis* controls pearl millet blast pathogen efficiently. The other experiment by Mahesh et al. (2019) reported effective control of pearl millet blast pathogen through *T. harzianum* strain Tri 5 and *P. fluorescens* strain RP-46. However, use of bio-agents in crops like pearl millet is very limited.

11.12.5 Use of Botanicals

Apart from host plant resistance, fungicides, and microbial biocontrol agents, plant products or extracts have been reported as antifungal agents against a wide range of pathogens (Amadioha 2000). The crude aqueous extracts or different commercial formulations of plants like neem, pongamia, derris, citrus grass, and tobacco have efficacy in managing diseases with less residual action and ecological safety. The commercial formulation of plant extracts like Agroneem (neem oil based herbal pesticide), raw neem oil (Azadirachtin), Soldier (*Aegle marmelos* @ 20%, *Ricinus communis* @ 20%, *Hygrophila spinosa* @ 20%, *Laminaria* spp. @ 20%, and *Lantana camara* @ 20%), Discheck (*Ficus benghalensis* @ 0.0001%, *Ficus religiosa* @ 0.0001%, *Ficus retusa* @ 0.0001%, Aqua solvent @ 99.99%), Neem gold (Azadirachtin @ 0.15%), and Nimbecidine (azadirachtin @ 0.03%) have been evaluated by Roopadevi and Patil (2017); Agroneem @ 1% was found effective in the management of blast pathogen of pearl millet. In another study, Kaurav et al. (2018) tested the efficacy of three botanicals like neem seed kernel extract (NSKE) @ 20%, aloe vera leaf extract @ 20%, and *Lantana camara* leaf extract @ 20% against *M. grisea* under *in vitro* conditions and found significant effect on the pathogen. Like bio-agents, botanicals too have little scope for their use as disease control agents in the crops like pearl millet at farm level. However, there is a need to test both botanicals and bio-agents as components of IDM package along with fungicides and host plant resistance for the management of pearl millet blast.

References

Adachi K, Hamer JE. Divergent cAMP signaling pathways regulate growth and pathogenesis in the rice blast fungus *Magnaporthe grisea*. Plant Cell. 1998;10(8):1361–73.

Agrios GN. Control of plant diseases in plant pathology. London: Academic; 1997. p. 200–16.

Ahn IP, Kim S, Kang S, Suh SC, Lee YH. Rice defense mechanisms against *Cochliobolus miyabeanus* and *Magnaporthe grisea* are distinct. Phytopathology. 2005;95(11):1248–55.

Amadioha AC. Controlling rice blast in vitro and in vivo with extracts of *Azadirachta indica*. Crop Prot. 2000;19(5):287–90.

Bao-Hua W, Guo-Dong L, Hai-Ming L, Yan L, Zong-Hua W. Assessment of *Magnaporthe grisea* mating type by PCR. Chin J Agric Biotechnol. 2005;2(1):1–5.

Barr ME. *Magnaporthe*, *Telimenella*, and *Hyponectria* (*Physosporellaceae*). Mycologia. 1977;69(5):952–66.

Bohnert HU, Fudal I, Dioh W, Tharreau D, Notteghem JL, Lebrun MH. A putative polyketide synthase/peptide synthetase from *Magnaporthe grisea* signals pathogen attack to resistant rice. Plant Cell. 2004;16(9):2499–513.

Borges-Walmsley MI, Walmsley AR. Triggers and targets of cAMP signalling-response. Trends Microbiol. 2000;8(7):302–3.

Bourett TM, Howard RJ. In vitro development of penetration structures in the rice blast fungus *Magnaporthe grisea*. Can J Bot. 1990;68(2):329–42.

Brindha VP, Viji G, Gnanamanickam SS. Mating type distribution, fertility and pathogenicity of rice isolates of *Magnaporthe grisea* in four rice growing regions of India. Indian Phytopathol. 1999;52(1):28–34.

Buckley TA, Allen BF. Notes on current investigations. Malays J Sustain Agric. 1951;34:133–41.

Cai J, Wang L, Pạn Q. Mating type and fertility of the rice blast fungus population derived from Guangdong province, China. Zhongguo Shuidao Kexue. 2005;38(4):837–42.

Cannon PF. The newly recognized family Magnaporthaceae and its interrelationships. Systema Ascomycetum. 1994;13(1):25–42.

Cavara F. Contributzioneallamicologialombarda. Atti Instituto Botanico Universita Laboratorio Crittogamico Pavia Ser. 1892;2:207–92.

Chandra Nayak S, Srivastava RK, Uday Shankar AC, Lavanya SM, Prakash G, Bishnoi HR, Kadvani DL, Singh OV, Niranjana SR, Prakash HS, Tara Satyavathi C. *Magnaporthe*: blast of pearl millet in India—present status and future prospects. All India Coordinated Research Project on Pearl Millet (Indian Council of Agricultural Research), Mandore, Jodhpur-342304. 2017. 51 p.

Choi W, Dean RA. The adenylate cyclase gene MAC1 of *Magnaporthe grisea* controls appressorium formation and other aspects of growth and development. Plant Cell. 1997;9(11):1973–83.

Consolo VF, Cordo CA, Salerno GL. Mating-type distribution and fertility status in *Magnaporthe grisea* populations from Argentina. Mycopathologia. 2005;160(4):285–90.

Coppin E, Debuchy R, Arnaise S, Picard M. Mating types and sexual development in filamentous ascomycetes. Microbiol Mol Biol Rev. 1997;61(4):411–28.

Couch BC, Kohn LM. A multilocus gene genealogy concordant with host preference indicates segregation of a new species, *Magnaporthe oryzae*, from *M. grisea*. Mycologia. 2002;94(4):683–93.

D'Souza CA, Heitman J. Conserved cAMP signaling cascades regulate fungal development and virulence. FEMS Microbiol Rev. 2001;25(3):349–64.

de Jong JC, McCormack BJ, Smirnoff N, Talbot NJ. Glycerol generates turgor in rice blast. Nature. 1997;389:244–5.

Dean RA. Signal pathways and appressorium morphogenesis. Annu Rev Phytopathol. 1997;35(1):211–34.

Dean RA, Talbot NJ, Ebbole DJ, Farman ML, Mitchell TK, Orbach MJ, Thon M, Kulkarni R, Xu JR, Pan H, Read ND. The genome sequence of the rice blast fungus *Magnaporthe grisea*. Nature. 2005;434(7036):980–6.

DeZwaan TM, Carroll AM, Valent B, Sweigard JA. *Magnaporthe grisea* pth11p is a novel plasma membrane protein that mediates appressorium differentiation in response to inductive substrate cues. Plant Cell. 1999;11(10):2013–30.

Dong-mei L, Hongli J, Wende L. Studies on mating type of *Magnaporthe grisea* in Sichuan Province of China. J Sichuan Agric Univ. 2005;23:168–74.

Ebbole DJ. *Magnaporthe* as a model for understanding host-pathogen interactions. Annu Rev Phytopathol. 2007;45:437–56.

Emechebe, AM. Some Aspects of Crop Diseases in Uganda. Makerere University Printery, Uganda. 1975.

Flaishman MA, Kolattukudy PE. Timing of fungal invasion using host's ripening hormone as a signal. Proc Natl Acad Sci U S A. 1994;91(14):6579–83.

Frankel OH, Brown AH. Plant genetic resources today: a critical appraisal. In: Holden JHW, Williams JT, editors. Crop genetic resources: conservation & evaluation. 1984.

Gilbert RD, Johnson AM, Dean RA. Chemical signals responsible for appressorium formation in the rice blast fungus *Magnaporthe grisea*. Physiol Mol Plant Pathol. 1996;48(5):335–46.

Gupta SK, Sharma R, Rai KN, Thakur RP. Inheritance of foliar blast resistance in pearl millet (*Pennisetum glaucum*). Plant Breed. 2012;131(1):217–9.

Hanna WW, Wells HD. Inheritance of Pyricularia leaf spot resistance in pearl millet. J Hered. 1989;80(2):145–7.

Hanna WW, Wells HD, Burton GW, Hill GM, Monson WG. Registration of 'Tifleaf 2' pearl millet. Crop Sci. 1988;28(6):1023.

Hebert TT. The perfect stage of *Pyricularia grisea*. Phytopathology. 1971;61(1):83–7.

Hebert TT. Production of the perfect stage of Pyricularia from rice and other hosts. Serie CE (CIAT) Centro Internacional de Agricultura Tropical. No. 9. 1975.

Howard RJ, Valent B. Breaking and entering: host penetration by the fungal rice blast pathogen *Magnaporthe grisea*. Annu Rev Microbiol. 1996;50(1):491–512.

Inukai T, Vales MI, Hori K, Sato K, Hayes PM. RMo 1 confers blast resistance in barley and is located within the complex of resistance genes containing Mla, a powdery mildew resistance gene. Mol Plant-Microbe Interact. 2006;19(9):1034–41.

Joosten MH, Cozijnsen TJ, De Wit PJ. Host resistance to a fungal tomato pathogen lost by a single base-pair change in an avirulence gene. Nature. 1994;367(6461):384–6.

Joshi HD, Gohel NM. Management of blast [*Pyricularia grisea* (Cooke) Sacc.] disease of pearl millet through fungicides. Bioscan. 2015;10(4):1855–8.

Kang S, Lee YH. Population structure and race variation of the rice blast fungus. Plant Pathol J. 2000;16:1–8.

Kang S, Chumley FG, Valent B. Isolation of the mating-type genes of the phytopathogenic fungus *Magnaporthe grisea* using genomic subtraction. Genetics. 1994;138(2):289–96.

Kankanala P, Czymmek K, Valent B. Roles for rice membrane dynamics and plasmodesmata during biotrophic invasion by the blast fungus. Plant Cell. 2007;19(2):706–24.

Kato H. Epidemiology of rice blast disease. J Plant Prot Res. 1974;7:1–20.

Kato H. Rice blast disease. Pestic Outlook. 2001;12(1):23–5.

Kato H, Yamaguchi T. The perfect state of *Pyricularia oryzae* Cav. from rice plants in culture. Jpn J Phytopathol. 1982;48(5):607–12.

Kato H, Mayama S, Sekine R, Kanazawa E, Izutani Y, Urashima AS, Kunoh H. Microconidium formation in *Magnaporthe grisea*. Jpn J Phytopathol. 1994;60(2):175–85.

Kaurav A, Pandya R, Singh B. Performance of botanicals and fungicides against blast of pearl millet (*Pennisetum glaucum*). Ann Plant Soil Res. 2018;20(3):258–62.

Kim CK, Kim CH. The rice leaf blast simulation model EPIBLAST. In: Systems approaches for agricultural development. Dordrecht: Springer; 1993. p. 309–21.

Kiran Babu T, Sharma R, Upadhyaya HD, Reddy PN, Senthilvel S, Sarma NDRK, Thakur RP. Evaluation of genetic diversity in *Magnaporthe grisea* populations adapted to finger millet using Simple Sequence Repeats (SSRs). Physiol Mol Plant Pathol. 2013;84:10–8.

Klaubauf S, Tharreau D, Fournier E, Groenewald JZ, Crous PW, De Vries RP, Lebrun MH. Resolving the polyphyletic nature of *Pyricularia* (Pyriculariaceae). Stud Mycol. 2014;79:85–120.

Kotasthane AS, Kumar J, Singh US, Zeigler RS. Genetics of fertility and mating type in *Magnaporthe grisea*. In: Rice blast: interaction with rice and control. Dordrecht: Springer; 2004. p. 261–70.

Kumar J, Nelson RJ, Zeigler RS. Population structure and dynamics of *Magnaporthe grisea* in the Indian Himalayas. Genetics. 1999;152(3):971–84.

Lamey HA. *Pyricularia oryzae* on rice seed in the United States. Plant Dis Rep. 1970;54(11):931–5.

Lanoiselet V, Cother E. Plant health Australia diagnostic protocol for *Magnaporthe grisea*. 2005. http://www.padil.gov.au/pbt | Rice Blast—*Magnaporthe grisea*.

Lavanya B, Gnanamanickam SS. Molecular tools for characterization of rice blast pathogen (*Magnaporthe grisea*) population and molecular marker-assisted breeding for disease resistance. Curr Sci. 2000;78:248–57.

Lee YH, Dean RA. Stage-specific gene expression during appressorium formation of *Magnaporthe grisea*. Exp Mycol. 1993a;17(3):215–22.

Lee YH, Dean RA. cAMP regulates infection structure formation in the plant pathogenic fungus *Magnaporthe grisea*. Plant Cell. 1993b;5(6):693–700.

Lee SC, Lee YH. Calcium/calmodulin-dependent signaling for appressorium formation in the plant pathogenic fungus *Magnaporthe grisea*. Mol Cells. 1998;8(6):698–704. Springer Science & Business Media BV.

Lee N, D'Souza CA, Kronstad JW. Of smuts, blasts, mildews, and blights: cAMP signaling in phytopathogenic fungi. Annu Rev Phytopathol. 2003;41(1):399–427.

Leslie JF, Klein KK. Female fertility and mating type effects on effective population size and evolution in filamentous fungi. Genetics. 1996;144(2):557–67.

Lubadde G, Tongoona P, Derera J, Sibiya J. Major pearl millet diseases and their effects on on-farm grain yield in Uganda. Afr J Agric Res. 2014;9(39):2911–8.

Lukose CM, Kadvani DL, Dangaria CJ. Efficacy of fungicides in controlling blast disease of pearl millet. Indian Phytopathol. 2011;60(1):137–8.

Mackill AO, Bonman JM. New hosts of *Pyricularia oryzae*. Plant Dis. 1986;70(2):125–7.

Mahesh M, Malatesha L, Venkataravana P. *In vitro* evaluation of fungicides and bioagents for the management of pearl millet blast caused by *Pyricularia grisea* (Cooke) Sacc. Int J Curr Microbiol Appl Sci. 2019;8(6):1422–9.

Manning K, Pelling R, Higham T, Schwenniger JL, Fuller DQ. 4500-year old domesticated pearl millet (*Pennisetum glaucum*) from the Tilemsi Valley, Mali: new insights into an alternative cereal domestication pathway. J Archaeol Sci. 2011;38(2):312–22.

Marley PS, Diourte M, Neya A, Nutsugah SK, Sereme P, Katile SO, Hess DE, Mbaye DF, Ngoko Z. Sorghum and pearl millet diseases in West and Central Africa. Sorghum Millets Dis. 2002;419–26.

Mehta PR, Singh B, Mathur SC. A new leaf spot disease of Bajra (*Pennisetum typhoides* Stapf and Hubbard) caused by a species of *Pyricularia*. Indian Phytopathol. 1953;5(2):142–3.

Mekwatanakarn P, Kositratana W, Phromraksa T, Zeigler RS. Sexually fertile *Magnaporthe grisea* rice pathogens in Thailand. Plant Dis. 1999;83(10):939–43.

Munch S, Lingner U, Floss DS, Ludwig N, Sauer N, Deising HB. The hemibiotrophic lifestyle of *Colletotrichum* species. J Plant Physiol. 2008;165(1):41–51.

Murakami J, Tosa Y, Kataoka T, Tomita R, Kawasaki J, Chuma I, Sesumi Y, Kusaba M, Nakayashiki H, Mayama S. Analysis of host species specificity of *Magnaporthe grisea* toward wheat using a genetic cross between isolates from wheat and foxtail millet. Phytopathology. 2000;90(10):1060–7.

Nagaraja A, Jagadish PS, Ashok EG, Gowda KK. Avoidance of finger millet blast by ideal sowing time and assessment of varietal performance under rain fed production situations in Karnataka. J Mycol Pathol Res. 2007;45(2):237–40.

Nakayama H, Nagamine T, Hayashi N. Genetic variation of blast resistance in foxtail millet (*Setaria italica* (L.) P. Beauv.) and its geographic distribution. Genet Resour Crop Evol. 2005;52(7):863–8.

Nishikado Y. Studies on rice blast disease. Bull. Bureau. Agric., Ministry of Agriculture and Forestry, Japan. 1926;15(1):211.

Nottéghem JL, Silue D. Distribution of the mating type alleles in *Magnaporthe grisea* populations pathogenic on rice. Phytopathology. 1992;82(4):421–4.

Oh Y, Donofrio N, Pan H, Coughlan S, Brown DE, Meng S, Mitchell T, Dean RA. Transcriptome analysis reveals new insight into appressorium formation and function in the rice blast fungus *Magnaporthe oryzae*. Genome Biol. 2008;9(5):85.

Ou SH. Pathogen variability and host resistance in rice blast disease. Annu Rev Phytopathol. 1980;18(1):167–87.

Ou SH. Rice diseases. Kew, Surrey: Commonwealth Mycological Institute; 1985.

Perfect SE, Green JR. Infection structures of biotrophic and hemibiotrophic fungal plant pathogens. Mol Plant Pathol. 2001;2(2):101–8.

Perrott RF, Chakraborty S. *Pyricularia grisea* causes blight of buffel grass (*Cenchrus ciliaris*) in Queensland, Australia. Trop Grassl. 1999;33(4):201–6.

Prabhu AS, Filippi MC, Castro N. Pathogenic variation among isolates of *Pyricularia oryzae* affecting rice, wheat, and grasses in Brazil. Int J Pest Manag. 1992;38(4):367–71.

Prakash G, Kumar A, Sheoran N, Aggarwal R, Satyavathi CT, Chikara SK, Ghosh A, Jain RK. First draft genome sequence of a pearl millet blast pathogen, *Magnaporthe grisea* Strain PMgDl, obtained using PacBio Single-Molecule Real-Time and Illumina NextSeq 500 Sequencing. Microbiol Resour Announc. 2019;8(20):e01499–18.

Priyadarshini VB, Viji G, Gnanamanickam SS. Mating type distribution, fertility and pathogenicity of rice isolates of *Magnaporthe grisea* in four rice growing regions of India. Indian Phytopathol. 1999;52(1):28–34.

Rai KN, Khairwal IS, Dangaria CJ, Singh AK, Rao AS. Seed parent breeding efficiency of three diverse cytoplasmic-nuclear male-sterility systems in pearl millet. Euphytica. 2009;165(3):495–507.

Roopadevi, Patil PV. *In-vitro* bioassay of fungicides, bioagents, botanicals against *Pyricularia grisea* (Cooke) Sacc.-Incitant of Pearl Millet Blast. Int J Pure Appl Biosci. 2017;5(4):1457–63.

Rossman AY, Howard RJ, Valent B. *Pyricularia grisea* the correct name for the rice blast disease fungus. Mycologia. 1990;82(4):509–12.

Saccardo PA. Fungorum extra-europaeorum Pugillus. Michelia. 1880;2(6):136–49.

Saikia UN, Puzari KC, Dutta PK. *Pennisetum pedicellatum*, a new host of *Pyricularia grisea*. Indian J Mycol Plant Pathol. 1982;12(2):237.

Sawada K. Blast of rice plants and its relation to the infective crops and weeds, with the description of five species of *Dactylaria*. Special Bull Taiwan Agric Exp Stat. 1917;16:1–78.

Sharma R, Upadhyaya HD, Manjunatha SV, Rai KN, Gupta SK, Thakur RP. Pathogenic variation in the pearl millet blast pathogen *Magnaporthe grisea* and identification of resistance to diverse pathotypes. Plant Dis. 2013;97(2):189–95.

Sharma R, Gate VL, Madhavan S. Evaluation of fungicides for the management of pearl millet [*Pennisetum glaucum* (L.)] blast caused by *Magnaporthe grisea*. Crop Prot. 2018;112:209–13.

Sharma R, Gupta, SK, Sharma S, Govindaraj M, Srivastava RK, Deshpande SP, Gupta R. Utilizing host plant resistance to manage blast: an emerging threat to dryland cereals. Lead paper presented at IPPC2019, 10–14 November, Hyderabad, India. 2019. p. 201.

Sharma R, Sharma S, Gate VL. Tapping *Pennisetum violaceum*, a wild relative of pearl millet (*Pennisetum glaucum*), for resistance to blast (caused by *Magnaporthe grisea*) and rust (caused by *Puccinia substriata* var. indica). Plant Dis. 2020;104:1487–91. https://doi.org/10.1094/PDIS-08-19-1602-RE.

Shaw BD, Kuo K, Hoch HC. Germination and appressorium development of *Phyllosticta ampelicida* pycnidiospores. Mycologia. 1998;90(2):258–68.

Shetty HS, Niranjan Raj S, Sudisha J, Thakur RP, Rai KN, Khairwal IS, Negi S, Mahala RS. *Pyricularia* blast of pearl millet in India. Crop Care. 2009;35:53–8.

Shirai M. Notes on plants collected in suruga, Totomi, yamato and kii. Bot Mag Tokyo. 1896;10:111–4.

Silue D, Notteghem JL. Production of perithecia of *Magnaporthe grisea* on rice plants. Mycol Res. 1990;94(8):1151–2.

Silva GB, Prabhu AS, Filippi MC, Trindade MG, Araújo LG, Zambolim L. Genetic and phenotypic diversity of *Magnaporthe oryzae* from leaves and panicles of rice in commercial fields in the State of Goias, Brazil. Trop Plant Pathol. 2009;34(2):77–86.

Singh DS, Pavgi MS. Perpetuation of *Pyricularia penniseti* causing brown leaf spot of bajra. Indian Phytopathol. 1977;30(2):242–4.

Singh S, Sharma R, Pushpavathi B, Gupta SK, Durgarani CV, Raj C. Inheritance and allelic relationship among gene (s) for blast resistance in pearl millet [*Pennisetum glaucum* (L.) R. Br.]. Plant Breed. 2018;137(4):573–84.

Skamnioti P, Gurr SJ. Against the grain: safeguarding rice from rice blast disease. Trends Biotechnol. 2009;27(3):141–50.

Sommerhalder RJ, McDonald BA, Zhan J. The frequencies and spatial distribution of mating types in *Stagonospora nodorum* are consistent with recurring sexual reproduction. Phytopathology. 2006;96(3):234–9.

Staples RC, Laccetti L, Yaniv Z. Appressorium formation and nuclear division in *Colletotrichum truncatum*. Arch Microbiol. 1976;109(1–2):75–84.

Steiner-Lange S, Fischer A, Boettcher A, Rouhara I, Liedgens H, Schmelzer E, Knogge W. Differential defense reactions in leaf tissues of barley in response to infection by *Rhynchosporium secalis* and to treatment with a fungal avirulence gene product. Mol Plant-Microbe Interact. 2003;16(10):893–902.

Suh JP, Roh JH, Cho YC, Han SS, Kim YG, Jena KK. The Pi40 gene for durable resistance to rice blast and molecular analysis of Pi40-advanced backcross breeding lines. Phytopathology. 2009;99(3):243–50.

Taguchi Y, Hyakumachi M, Horinouchi H, Kawane F. Biological control of rice blast disease by *Bacillus subtilis* IK-1080. Jpn J Phytopathol. 2003;69(2):85–93.

Takan JP, Chipili J, Muthumeenakshi S, Talbot NJ, Manyasa EO, Bandyopadhyay R, Sere Y, Nutsugah SK, Talhinhas P, Hossain M, Brown AE. *Magnaporthe oryzae* populations adapted to finger millet and rice exhibit distinctive patterns of genetic diversity, sexuality and host interaction. Mol Biotechnol. 2012;50(2):145–58.

Talbot NJ. On the trail of a cereal killer: exploring the biology of *Magnaporthe grisea*. Annu Rev Microbiol. 2003;57(1):177–202.

TeBeest DO, Guerber C, Ditmore M. Rice blast. The plant health instructor. 2007. https://doi.org/10.1094/PHI-I-2007-0313-07.

Teng PS. Epidemiological basis for blast management. In: Zeigler RS, Leong SA, Teng PS, editors. Rice blast disease. Wallingford: CAB International and IRRI; 1994. p. 409–33.

Thakur RP, Sharma R, Rai KN, Gupta SK, Rao VP. Screening techniques and resistance sources for foliar blast in pearl millet. J SAT Agric Res. 2009;7:1–5.

Thakur RP, Sharma R, Rao VP. Screening techniques for pearl millet diseases. Information Bulletin No. 89. International Crops Research Institute for the Semi-Arid Tropics, Patancheru 502 324, Telangana, India. 2011. p. 48.

Thieringer R, Kunau WH. The beta-oxidation system in catalase-free microbodies of the filamentous fungus *Neurospora crassa*. Purification of a multifunctional protein possessing 2-enoyl-CoA hydratase, L-3-hydroxyacyl-CoA dehydrogenase, and 3-hydroxyacyl-CoA epimerase activities. J Biol Chem. 1991;266(20):13110–7.

Thines E, Weber RW, Talbot NJ. MAP kinase and protein kinase A-dependent mobilization of triacylglycerol and glycogen during appressorium turgor generation by *Magnaporthe grisea*. Plant Cell. 2000;12(9):1703–18.

Timper P, Wilson JP, Johnson AW, Hanna WW. Evaluation of pearl millet grain hybrids for resistance to *Meloidogyne* spp. and leaf blight caused by *Pyricularia grisea*. Plant Dis. 2002;86(8):909–14.

Tredway LP, Stevenson KL, Burpee LL. Mating type distribution and fertility status in *Magnaporthe grisea* populations from turf grasses in Georgia. Plant Dis. 2003;87(4):435–41.

Ueyama A, Tsuda M. Formation of the perfect state in culture of *Pyricularia* sp. from some graminaceous plants Preliminary report. Trans Mycol Soc Japan. 1975.

Upadhyaya HD, Ortiz R. A mini core subset for capturing diversity and promoting utilization of chickpea genetic resources in crop improvement. Theor Appl Genet. 2001;102(8):1292–8.

Urak RZ, Douhan GW, Wong F. Mating-type distribution of the rice blast pathogen *Pyricularia grisea* in California. UCR Undergrad Res J. 2008;2:61–5.

Werder J, Manzo SK. Pearl millet diseases in Western Africa. Sorghum and millets diseases: a second world review. Patancheru: ICRISAT 109-4; 1992.

Wilson JP, Gates RN. Forage yield losses in hybrid pearl millet due to leaf blight caused primarily by *Pyricularia grisea*. Phytopathology. 1993;83(7):739–44.

Wilson JP, Hanna WW. Disease resistance in wild *Pennisetum* species. Plant Dis. 1992a;76(11):1171–5.

Wilson JP, Hanna WW. Effects of gene and cytoplasm substitutions in pearl millet on leaf blight epidemics and infection by *Pyricularia grisea*. Phytopathology. 1992b;82(8):839–42.

Wilson RA, Talbot NJ. Under pressure: investigating the biology of plant infection by *Magnaporthe oryzae*. Nat Rev Microbiol. 2009;7(3):185–95.

Wilson JP, Wells HD, Burton GW. Inheritance of resistance to *Pyricularia grisea* in pearl millet accessions from Burkina Faso and inbred Tift 85DB. J Hered. 1989;80(6):499–501.
Xu JR. MAP kinases in fungal pathogens. Fungal Genet Biol. 2000;31(3):137–52.
Xu JR, Hamer JE. Assessment of *Magnaporthe grisea* mating type by spore PCR. Fungal Genet Rep. 1995;42(1):80.
Xu JR, Hamer JE. MAP kinase and cAMP signaling regulate infection structure formation and pathogenic growth in the rice blast fungus *Magnaporthe grisea*. Genes Dev. 1996;10(21):2696–706.
Yaegashi H. On the sexuality of blast fungi, *Pyricularia* spp. Jpn J Phytopathol. 1977;43(4):432–9.
Yaegashi H, Hebert TT. Perithecial development and nuclear behavior in *Pyricularia*. Phytopathology. 1976;66:122–6.
Yaegashi H, Yamada M. Pathogenic race and mating type of *Pyricularia oryzae* from Soviet Union, China, Nepal, Thailand, Indonesia and Colombia. Jpn J Phytopathol. 1986;52(2):225–34.
Yella Goud T. Epidemiology, virulence and molecular diversity in blast [*Magnaporthe grisea* (Hebert) Barr.] of pearl millet [*Pennisetum glaucum* (L.) R. Br.] and resistance in the host to diverse pathotypes. Doctoral dissertation, Professor Jayashankar Telangana State Agricultural University, Hyderabad. 2015.
Yella Goud T, Sharma R, Gupta SK, Devi GU, Gate VL, Boratkar M. Evaluation of designated hybrid seed parents of pearl millet for blast resistance. Indian J Plant Prot. 2016;44(1):83–7.
Yoder OC, Valent B, Chumley F. Genetic nomenclature and practice for plant pathogenic fungi. Phytopathology. 1986;76(3):383–5.
Zarrinnia V, Nikkhah MJ, Zadeh HR, Bitarafan F, Sedaghatfar E. Fertility assessment and mating type analysis reveal the absence of sexual stage in *Magnaporthe oryzae* populations in Iran. Sci Series Data Rep. 2012;4:123–32.
Zeigler RS. Recombination in *Magnaporthe grisea*. Annu Rev Phytopathol. 1998;36(1):249–75.
Zeng J, Feng S, Cai J, Wang L, Lin F, Pan Q. Distribution of mating type and sexual status in Chinese rice blast populations. Plant Dis. 2009;93(3):238–42.
Zheng Y, Zhang G, Lin F, Wang Z, Jin G, Yang L, Wang Y, Chen X, Xu Z, Zhao X, Wang H. Development of microsatellite markers and construction of genetic map in rice blast pathogen *Magnaporthe grisea*. Fungal Genet Biol. 2008;45(10):1340–7.

Chapter 12
Advances in Genetics and Genomics for Management of Blast Disease in Cereal Crops

B. N. Devanna, Rajashekara Hosahatti, S. Raghu, P. K. Singh, P. Jain, C. Parameswaran, J. Singh, K. Arora, S. Samantaray, J. L. Katara, and R. L. Verma

Contents

12.1 Introduction

The hemibiotrophic fungal pathogen of genus *Pyricularia*, a filamentous fungus, infects monocot plants leading to blast disease and this genus is separated into multiple groups on the basis of conidial morphology, phylogeny, and mating compatibility (Kato et al. 2000; Hirata et al. 2007; Tosa and Chuma 2014; Murata et al. 2014; Klaubauf et al. 2014). Among them, *Magnaporthe oryzae* forms the largest cluster and isolates from this infect cereal crops like rice and wheat, and isolates from second biggest cluster (*P. grisea*) infect crabgrass (*Digitaria sanguinalis*)

B. N. Devanna (✉) · S. Raghu · C. Parameswaran · S. Samantaray · J. L. Katara
R. L. Verma
ICAR-National Rice Research Institute, Cuttack, Odisha, India

R. Hosahatti
ICAR-Vivekananda Parvatiya Krishi Anusandhan Sansthan, Almora, Uttarakhand, India

P. K. Singh
National Agri-Biotechnology Institute, Mohali, Punjab, India

P. Jain · J. Singh · K. Arora
ICAR-National Institute for Plant Biotechnology, New Delhi, India

S. C. Nayaka et al. (eds.), *Blast Disease of Cereal Crops*, Fungal Biology,
https://doi.org/10.1007/978-3-030-60585-8_12

(Hirata et al. 2007). Isolates infecting grasses, such as *Leersia/Setaria* and *Cenchrus/Echinochloa*, are labelled as *Pyricularia* sp. (L/S) and (C/E), respectively.

This classification of *Magnaporthe* isolates based on the host species they infect needs further understanding under the light of recent studies. The previous host specificity studies of pathotypes of *Magnaporthe* had limitation with respect to the number of pathogen isolates and host genotypes. However, for example the recent studies involving MoT strains reveal that besides wheat, they are also pathogenic on oat, barley and rye, but non-virulent on rice (Urashima et al. 1993). Similarly, *M. oryzae* isolates from rice (MoO) are reported to be infecting barley (Chen et al. 2003; Zellerhoff et al. 2006; Hyon et al. 2012). However, with some exceptions, barley is a non-host to isolates of *Magnaporthe* sp. which are closely related to *M. grisea* of fountain grass (*Pennisetum*) and crabgrass (*Digitaria sanguinalis*) (Zellerhoff et al. 2006; Hyon et al. 2012).

The availability of genomic resources in these host plants and *Magnaporthe* species has been effectively used to characterise the pathogen isolates. Further, sequence information of genomes of host plants such as rice, wheat, barley and millets further facilitated the understanding of host-pathogen interaction. Rice-*M. oryzae* is now considered as a model system for understanding plant-pathogen interaction in crop plants (Sharma et al. 2012).

12.2 Genetics of Blast Disease Resistance

Blast resistance genes were initially described in 1923 by Sasaki, and later it was further elaborated by Takahasihi (1965) and Yamasaki and Kiyosawa (1966). Systematic reports on rice blast resistance genes were initially undertaken by Goto et al. (1964), who established the differential system for blast fungus races in Japan. The first blast resistance gene *Pi-a* was identified from japonica variety Aichi Asahi by Kiyosawa (1967). Majority of the studies have shown that resistance to blast disease is dominant and is governed by single gene. The inheritance pattern of disease resistance is easily traced in F_2 mapping population specifically developed to identify associated blast genes/QTLs (Mackill and Bonman 1992). The virulence in *M. oryzae* is subjected to the selection pressure created by resistant genotypes. The high rate of natural mutations due to selection pressure created by resistance genes has led to accumulation of more virulent genes. Thus, cultivars carrying a single resistance gene are more prone to breakdown of resistance leading to short lifespan of resistant genotypes in the field due to higher variation in *M. oryzae Avr* genes (Bonman et al. 1989; Valent and Chumley 1994). The interaction between plant *R* gene and pathogen *Avr* gene is very dynamic and resistance response is largely mediated by a pair of one or two genes (Notteghem et al. 1994; Yu et al. 1996; Pan et al. 1998; Chen et al. 2005; Padmavathi et al. 2005; Sharma et al. 2007). Therefore, it is imperative to identify a new source of *R* gene for gene deployment either alone or through pyramiding of multiple and divergent genes from different sources

(Padmavathi et al. 2005; Sharma et al. 2012; Devanna et al. 2014; Kumari et al. 2017, 2018). The most popular work on blast resistance was carried out by Kiyosowa and his colleagues in Japan, wherein they were able to identify as many as 14 alleles at eight resistance loci (Kiyosawa and Ando 1990).

Allelism tests would be much easier in NILs than in donor cultivars where resistance is often conferred by two or more major genes. The genetic mechanism of broad-spectrum resistance to blast pathogen in rice landraces was reported using the standard blast differential system comprising the standard isolates and differential rice lines for analysing segregation ratio governing either dominant or recessive genes (Koide et al. 2013). The inheritance pattern of resistance in landrace with different potential *R* genes could be useful to know the nature of gene action (Rajashekara et al. 2014).

12.3 Resistance Genes for Blast Resistance in Crops

The mechanism of blast disease resistance follows the classical model of gene-for-gene hypothesis (Silue et al. 1992), wherein plant R-protein recognises the cognate AVR-protein. *Magnaporthe* produces an array of effector proteins, including those associated with avirulence activity, which are subsequently recognised by R-proteins of resistance plants resulting in the activation of innate immunity. The deployment of *R* gene in popular varieties of crop plants for blast resistance is the economically viable and eco-friendly approach. Resistance is largely conferred to either single *R* gene which imparts complete immunity towards few or multiple strains or minor genes/QTLs which confer partial resistance. Major blast resistance genes with broad-spectrum resistance have been useful in providing durable resistance; however, in the light of rapidly evolving *Magnaporthe* strains, deployment of cautious and scientifically *R* genes with overlapping resistance pattern can effectively delay the breakdown in resistance reaction (Chen et al. 2005). The existing genetic diversity of the majority crops is largely preserved in the form of collection of germplasm/accessions. However, the diversity of these gene pools for disease resistance has not been fully exploited for resistance genes and alleles (Couch et al. 2002).

Large number of resistance genes encode proteins having nucleotide-binding site and leucine-rich repeat (NBS-LRR) domains. NBS domains having conserved motifs are characterised to bind and hydrolyse ATP and GTP, whereas the LRR motif is typically responsible for protein–protein interactions. These NBS–LRR class of resistance proteins play an important role in the resistance against *Magnaporthe* infection. These NBS–LRR proteins interact either directly with pathogen effectors through the gene-for-gene hypothesis or through other host proteins (guard hypothesis) (Van der Biezen and Jones 1998). Till now, more than 100 *R* genes and around 350 QTLs for blast resistance have been reported in rice, and 27 of these *R* genes have been cloned and functionally characterised (Rajashekara et al. 2014; Su et al. 2015; Zheng et al. 2016; Zhu et al. 2016; Devanna and Sharma 2018; Singh et al. 2020).

12.4 Genomic Resources Developed for Enhancing Blast Disease Resistance in Crop Plants

Management of blast disease in rice and other crops has been highly facilitated and speeded up with the availability of genomics resources, both in crops and blast pathogen. Though *Magnaporthe* infects almost 50 grass species, genome sequence information is available for commonly cultivated crops like rice, wheat, millets, Brachypodium and barley among the others (https://www.ncbi.nlm.nih.gov/genome/browse/#!/overview/). These genomic resources have largely enabled the researchers in deciphering the plant pathogen interaction during the process of blast pathogen infection, mapping and map-based cloning of candidate resistance genes, genome-wide studies for identification of putative *R* genes and mining for novel alleles of major blast resistance genes (Inukai et al. 2006; Sharma et al. 2012; Kumari et al. 2013; Devanna and Sharma 2018). The major rice blast resistance gene *Pi54* was the first *R* gene to be mapped and subsequently cloned using the sequence information of rice genome (Sharma et al. 2005; Rai et al. 2011). To date more than 100 QTLs for blast resistance have been mapped and around 25 of them have been cloned and characterised in rice (Devanna and Sharma 2018). Unlike rice, *R* genes for blast resistance are comparatively uncommon in wheat (Cruz and Valent 2017). In wheat, eight blast resistance genes have been mapped to different chromosomes using molecular marker information and recent wheat genomic resources (Devanna and Sharma 2018). However, to the best of our knowledge, no blast resistance gene has been cloned from wheat. This may have been attributed to the complex nature of wheat genome, and also to some extent the efficiency of genetic transformation of wheat plant for functional analysis of cloned genes (Jasdeep and Avijit 2016).

In other cereal crop, barley, around 16 QTLs and 2 genes for blast resistance were mapped using available genomics resources and marker-based comparative analysis of rice and barley genomic locations (Sato et al. 2001; Chen et al. 2003; Inukai et al. 2006; Nga et al. 2012) (Table 12.1). In case of finger millet also, around nine QTLs have been mapped using the molecular markers (Babu et al. 2014; Ramakrishnan et al. 2016). Though the other millet, pearl millet, has its genome sequenced and blast is a major disease affecting its yield, not much headway has been made in the identification of resistance resources for gene/QTL mapping. Recently a QTL (R-QTL) was mapped to chromosome 1 of pearl millet using SSR marker information (Sanghani et al. 2018).

Table 12.1 List of gene/QTLs identified in barley for blast disease resistance in barley

Gene/QTLs	Resistance source	Chromosome	Reference
RMo1	Baronesse	1H	Inukai et al. (2006)
Rmo2	–	7H	Nga et al. (2012)
Four QTLs	TR306	3H;4H;7H	Sato et al. (2001)
bbr2H, bbr3H, bbr4Ha, bbr4Hb, bbr5Hc, bbr5Hd, bbr7Ha, bbr7Hb, bb7Hc	TR306	2H,3H,4H,5H,7H	Chen et al. (2003)
bbr5Ha, bbr5Hb, bbr6H	Harrington	6H,6H	Chen et al. (2003)

Genome sequence information has also facilitated genome-wide study of resistance genes and their characterisation. Rice has around 400 predicted NBS–LRR genes (McHale et al. 2006). On the contrary, the hexaploid wheat has over 2000 NBS–LRR genes (Gu et al. 2015). These genomic resources would facilitate the identification and characterisation of novel blast resistance genes in crop plants.

12.5 Genomic Resources in *Magnaporthe*

The first genome sequence information for the blast fungus *M. grisea* was available by 2005 (Dean et al. 2005). This study highlighted the genomic composition coding for a large array of secretory proteins, and presence of high copy number of transposable elements indicating the clonal nature of blast fungus. Later, two field isolates of *M. oryzae* from India were sequenced, and this study unravelled transposable element-mediated genomic variations in the form of SNPs, InDels and ICVs and also predicted the Indian isolate-specific novel genes (Gowda et al. 2015). Till date around 21 isolates of *Magnaporthe* sp. infecting different host plants have been sequenced (Table 12.2). These studies have revealed traces of introgressed genomes from non-rice isolates into rice isolate (Chiapello et al. 2015). It is also found that the host specificity of *Magnaporthe oryzae* isolates is related with a divergence between lineages but without any major gene flow, and in rice the adaptation of *Oryza* isolates was linked with the presence of isolate-specific gene families, though in small numbers.

The availability of genomic resources has facilitated the understanding of molecular aspects of blast disease resistance in crops, and also the mechanism of pathogenesis in *Magnaporthe*. The genome sequence information of blast fungus along with their evolutionary relation and identification of avirulence (*Avr*) genes will help in understanding their host-plant relationship and also facilitate the decision-making process for the effective deployment of resistance genes in host plants (Singh et al. 2019). Identification and characterisation of *Avr*-genes also help in better understanding the mechanism infection by the pathogen, its target host proteins and pathogenicity pattern of the fungal population. Though genomes of *Magnaporthe* infecting different hosts have been available, information regarding cloned *Avr*-genes is available only in rice isolates. Till today, around 13 *Avr*-genes have been cloned in *M. oryzae* isolates from rice (Sharma et al. 2012; Ray et al. 2016; Lopez et al. 2019). Besides the genome sequence information, the transcriptomic analysis of host plant during the infection process has also been reported in major cereal crops such as rice and wheat (Tufan et al. 2009; Sharma et al. 2016; Islam et al. 2016). The interested readers can consult the detailed review on transcriptomic analysis rice during rice-*M. oryzae* interaction (Sharma et al. 2016). Therefore, the genomics resources generated in the *Magnaporthe* isolates and their host plants are increasing with the development in the field of genome sequencing techniques, and these resources are being effectively used for the identification and characterisation of novel genes.

Table 12.2 Sequence information of blast fungus *Magnaporthe* from crop plants

Strain/isolate	Host plant	Genome size (Mb)	Country	Reference
70–15	Rice	38.8	USA	Dean et al. (2005)
Ina168	Rice	40.48	Japan	Yoshida et al. (2009)
P131 and Y34	Rice	37.6 and 38.2, respectively	Japan and China, respectively	Xue et al. (2012)
FJ81278 and HN19311	Rice	37.3 and 37.1, respectively	China	Chen et al. (2013)
B157 and MG01	Rice	41 and 43, respectively	India	Gowda et al. (2015)
FR13	Rice	42.4	France	Chiapello et al. (2015)
PH14	Rice	40	Philippines	Chiapello et al. (2015)
TH16	Rice	39.1	Thailand	Chiapello et al. (2015)
TH12	Rice	40.1	Thailand	Chiapello et al. (2015)
Guy11	Rice	37.49	French Guyana	Bao et al. (2017)
RP-2421	Rice	44.85	India	Singh et al. (2019)
BR32	Wheat	41.9	Brazil	Chiapello et al. (2015)
B71	Wheat	43	Bolivia	Peng et al. (2019)
BTJP4-1	Wheat	44.56	Bangladesh	Win et al. (2019)
BTMP13-1	Wheat	43.97	Bangladesh	Win et al. (2019)
BTGP1-b	Wheat	44.40	Bangladesh	Win et al. (2019)
BTGP6-f	Wheat	44.23	Bangladesh	Win et al. (2019)
US71	Foxtail millet	41.2	United States	Chiapello et al. (2015)
CD156	Goosegrass	42.7	Ivory Coast	Chiapello et al. (2015)
BR29	Crabgrass	40.9	Brazil	Chiapello et al. (2015)
PMg_Dl	Pearl millet	47.90	India	Prakash et al. (2019)

References

Babu BK, Dinesh P, Agrawal PK, Sood S, Chandrashekara C, Bhatt JC. Comparative genomics and association mapping approaches for blast resistant genes in finger millet using SSRs. PLoS One. 2014;9(6):e99182. https://doi.org/10.1371/journal.pone.0099182.

Bao J, Chen M, Zhong Z. PacBio sequencing reveals transposable elements as a key contributor to genomic plasticity and virulence variation in *Magnaporthe oryzae*. Mol Plant. 2017;10(11):1465–8. https://doi.org/10.1016/j.molp.2017.08.008.

Bonman JM, Bandong JM, Lee YH, Lee E, Valent B. Race-specific partial resistance to blast in temperate japonica rice cultivars. Plant Dis. 1989;73:496–9.

Chen H, Wang S, Xing Y, Xu C, Hayes PM, Zhang Q. Comparative analyses of genomic locations and race specificities of loci for quantitative resistance to *Pyricularia grisea* in rice and barley. Proc Natl Acad Sci. 2003;100(5):2544–9. https://doi.org/10.1073/pnas.0437898100.

Chen S, Wang L, Que Z, Pan R, Pan Q. Genetic and physical mapping of Pi37 (t), a new gene conferring resistance to rice blast in the famous cultivar St. No. 1. Theor Appl Genet. 2005;111:1563–70.

Chen C, Lian B, Hu J. Genome comparison of two *Magnaporthe oryzae* field isolates reveals genome variations and potential virulence effectors. BMC Genomics. 2013;14:887. https://doi.org/10.1186/1471-2164-14-887.

Chiapello H, Mallet L, Guérin C, Aguileta G, Amselem J, Kroj T, Ortega-Abboud E, Lebrun M-H, Henrissat B, Gendrault A, Rodolphe F, Tharreau D, Fournier E. Deciphering genome content and evolutionary relationships of isolates from the fungus *Magnaporthe oryzae* attacking different host plants. Genome Biol Evol. 2015;7(10):2896–912. https://doi.org/10.1093/gbe/evv187.

Dean R, Talbot N, Ebbole D. The genome sequence of the rice blast fungus *Magnaporthe grisea*. Nature. 2005;434:980–6.

Devanna BN, Sharma TR. Wheat blast disease management: cues from the advancements in molecular biology of rice-*Magnaporthe* pathosystem. J Plant Biochem Biotechnol. 2018;27:249–59.

Devanna NB, Vijayan J, Sharma TR. The blast resistance gene *Pi54of* cloned from *Oryza officinalis* interacts with *Avr-Pi54* through its novel non-LRR domains. PLoS One. 2014;9(8):e104840. https://doi.org/10.1371/journal.pone.0104840.

Goto K, et al. Ditto, 2. Special Report on Forecasting of Occurrence of Disease and Insect Pest, No. 18. 1964. p. 132.

Gowda M, Shirke MD, Mahesh HB, Chandarana P, Rajamani A, Chattoo BB. Genome analysis of rice-blast fungus Magnaporthe oryzae field isolates from southern India. Genomics Data. 2015;5:284–91. https://doi.org/10.1016/j.gdata.2015.06.018. Published 2015 Jun 20.

Gu L, Si W, Zhao L, Yang S, Zhang X. Dynamic evolution of NBS-LRR genes in bread wheat and its progenitors. Mol Gen Genomics. 2015;290(2):727–38. https://doi.org/10.1007/s00438-014-0948-8.

Hirata K, Kusaba M, Chuma I, Osue J, Nakayashiki H, Mayama S, Tosa Y. Speciation in Pyriculariainferred from multilocus phylogenetic analysis. Mycol Res. 2007;111:799–808.

Hyon G-S, Nga N, Chuma I, Inoue Y, Asano H, Murata N, Kusaba M, Tosa Y. Characterization of interactions between barley and various host-specific subgroups of *Magnaporthe oryzae* and *M. grisea*. J Gen Plant Pathol. 2012;78(4):237–46.

Inukai T, Vales MI, Hori K, Sato K, Hayes PM. RMo1 confers blast resistance in barley and is located within the complex of resistance genes containing Mla, a powdery mildew resistance gene. Mol Plant-Microbe Interact. 2006;19(9):1034–41. https://doi.org/10.1094/MPMI-19-1034.

Islam MT, Croll D, Gladieux P. Emergence of wheat blast in Bangladesh was caused by a South American lineage of *Magnaporthe oryzae*. BMC Biol. 2016;14(1):84. https://doi.org/10.1186/s12915-016-0309-7.

Jasdeep CP, Avijit T. Genetic transformation and transgenic wheat development: an overview. Cloning Transgenesis. 2016;5:147. https://doi.org/10.4172/2168-9849.1000147.

Win J, Chanclud E, Reyes-Avila CS, Langner T, Islam T, Kamoun S. Nanopore sequencing of genomic DNA from Magnaporthe oryzae isolates from different hosts. Zenodo. (2019) https://doi.org/10.5281/zenodo.2564950.

Kato H, Yamamoto M, Yamaguchi-Ozaki T, Kadouchi H, Iwamoto Y, Nakayashiki H, Tosa Y, Mayama S, Mori N. Pathogenicity, mating ability and DNA restriction fragment length polymorphisms of Pyricularia populations isolated from Gramineae, Bambusideae and Zingiberaceaeplants. J Plant Pathol. 2000;66:30–47.

Kiyosawa S. Genetic studies on host-pathogen relationship in the rice blast disease. In: Proc. Symp. Rice diseases and their control by growing resistant varieties and other measures, Tokyo. 1967. p. 137–53.

Kiyosawa S, Ando I. Blast resistance. In: Matzo T, editor. Science of rice plant. Vol. 3. Genetics. Tokyo: Nosan-Gyoson Bunka Kyokai; 1990. p. 361–85.

Kumari A, Das A, Devanna BN, Thakur SS, Singh PK, Singh NK, Sharma TR. Mining of rice blast resistance gene Pi54 shows effect of single nucleotide polymorphisms on phenotypic expression of the alleles. Eur J Plant Pathol. 2013;137:55–65. https://doi.org/10.1007/s10658-013-0216-5.

Kumari M, Rai AK, Devanna BN. Co-transformation mediated stacking of blast resistance genes *Pi54* and *Pi54rh* in rice provides broad spectrum resistance against *Magnaporthe oryzae*. Plant Cell Rep. 2017;36(11):1747–55. https://doi.org/10.1007/s00299-017-2189-x.

Kumari M, Devanna BN, Singh PK, Rajashekara H, Sharma V, Sharma TR. Stacking of blast resistance orthologue genes in susceptible indica rice line improves resistance against *Magnaporthe oryzae*. 3 Biotech. 2018;8(1):37. https://doi.org/10.1007/s13205-017-1062-5.

Lopez ALC, Yli-Matilla T, Cumagun CJR. Geographic distribution of avirulence genes of the rice blast fungus *Magnaporthe oryzae* in the Philippines. Microorganisms. 2019;7(1):23. https://doi.org/10.3390/microorganisms7010023.

Mackill DJ, Bonman JM. Inheritance of blast resistance in near-isogenic lines of rice. Phytopathology. 1992;82:746–9.

Murata N, Aoki T, Kusaba M, Tosa Y, Chuma I. Various species of *Pyricularia* constitute a robust clade distinct from *Magnaporthe salvinii* and its relatives in Magnaporthaceae. J Gen Plant Pathol. 2014;80:66–72.

Nga NTT, Inoue Y, Chuma I, Hyon G-S, Okada K, Vy TTP, Kusaba M, Tosa Y. Identification of a novel locus Rmo2 conditioning resistance in barley to host-specific subgroups of *Magnaporthe oryzae*. Phytopathology. 2012;102:674–82.

Notteghem JL, Sillue DD, Roumen E. Present knowledge of rice resistance and strategies for *Magnaporthe grisea* pathogenicity and avirulence gene analysis. Rice blast disease. Wallingford, Oxfordshire: CAB International; 1994. p. 155–65.

Padmavathi G, Ram T, Satyanarayana K, Mishra B. Identification of blast (*Magnaporthe grisea*) resistance genes in rice. Curr Sci. 2005;88:628–30.

Pan QH, Wang L, Tanisaka T, Ikehashi H. Allelism of rice blast resistance genes in two Chinese native cultivars and identification of two new resistance genes. Plant Pathol. 1998;47:165–70.

Peng Z, Oliveira-Garcia E, Lin G, Hu Y, Dalby M, Migeon P, et al. Effector gene reshuffling involves dispensable mini-chromosomes in the wheat blast fungus. PLoS Genet. 2019;15(9):e1008272. https://doi.org/10.1371/journal.pgen.1008272.

Prakash G, Kumar A, Sheoran N, et al. First draft genome sequence of a pearl millet blast pathogen, *Magnaporthe grisea* strain PMg_Dl, obtained using PacBio single-molecule real-time and Illumina NextSeq 500 sequencing. Microbiol Resour Announc. 2019;8(20):e01499–18. https://doi.org/10.1128/MRA.01499-18. Published 2019 May 16.

Rai AK, Kumar SP, Gupta SK, Gautam N, Singh NK, Sharma TR. Functional complementation of rice blast resistance gene Pi-k ^{h} (Pi54) conferring resistance to diverse strains of *Magnaporthe oryzae*. J Plant Biochem Biotechnol. 2011;20:55–65. https://doi.org/10.1007/s13562-010-0026-1.

Rajashekara H, Ellur RK, Khanna A, Nagarajan M, Krishnan SG, Singh AK, Sharma P, Sharma TR, Singh UD. Inheritance of blast resistance and its allelic relationship with five major R genes in a rice landrace "Vanasurya". Indian Phytopathol. 2014;67(4):365–9.

Ramakrishnan M, Antony Ceasar S, Duraipandiyan V, Vinod KK, Kalpana K, Al-Dhabi NA, et al. Tracing QTLs for leaf blast resistance and agronomic performance of finger millet (*Eleusine coracana* (L.) Gaertn.) genotypes through association mapping and *in silico* comparative genomics analyses. PLoS One. 2016;11(7):e0159264. https://doi.org/10.1371/journal.pone.0159264.

Ray S, Singh PK, Gupta DK, et al. Analysis of *Magnaporthe oryzae* genome reveals a fungal effector, which is able to induce resistance response in transgenic rice line containing resistance gene, Pi54. Front Plant Sci. 2016;7:1140. https://doi.org/10.3389/fpls.2016.01140.

Sanghani BK, Sanghani AO, Kothari VV, Raval SS, Kahodariya JH, Ramani HR, Vadher KJ, Gajera HP, Golakiya BA, Mandavia MK. The SSR based linkage map construction and identification of QTLs for blast (*Pyricularia grisea*) resistance in pearl millet (*Pennisetum glaucum* (L.) r. br.). J Pharmacogn Phytochem. 2018;7(2):3057–64.

Sato K, Inukai T, Hayes P. QTL analysis of resistance to the rice blast pathogen in barley (*Hordeum vulgare*). Theor Appl Genet. 2001;102:916–20.

Sharma TR, Madhav MS, Singh BK, Shanker P, Jana TK, Dalal V, Pandit A, Singh A, Gaikwad K, Upreti HC, Singh NK. High resolution mapping, cloning and molecular characterization of the Pi-k ^{h} gene of rice, which confers resistance to *M. grisea*. Mol Gen Genomics. 2005;274:569–78.

Sharma RC, Shrestha SM, Pandey MP. Inheritance of blast resistance and associated microsatellite markers in rice cultivar 'Laxmi'. J Phytopathol. 2007;155:749–53.

Sharma TR, Rai AK, Gupta SK, Vijayan J, Devanna BN, et al. Rice blast management through host-plant resistance: retrospect and prospects. Agric Res. 2012;1:37–52. https://doi.org/10.1007/s40003-011-0003-5.

Sharma TR, Das A, Thakur S, Devanna BN, Singh PK, Jain P, Vijayan J, Kumar S. Oscillating transcriptome during rice-Magnaporthe interaction. Curr Issues Mol Biol. 2016;19:99–120.

Silue DJ, Notteghem L, Tharreau D. Evidence of gene for-gene relationship in the *Oryza sativa Magnaporthe grisea* pathosystem. Phytopathology. 1992;82:577–80.
Singh PK, Mahato AK, Jain P, Rathour R, Sharma V, Sharma TR. Comparative genomics reveals the high copy number variation of a retro transposon in different *Magnaporthe* isolates. Front Microbiol. 2019;10:966. https://doi.org/10.3389/fmicb.2019.00966.
Singh J, Gupta SK, Devanna BN, Singh S, Upadhyay A, Sharma TR. Blast resistance gene Pi54 over-expressed in rice to understand its cellular and sub-cellular localization and response to different pathogens. Sci Rep. 2020;10:5243. https://doi.org/10.1038/s41598-020-59027-x.
Takahasihi Y. Genetics of resistance to the rice blast disease. The rice blast disease. Baltimore: Johns Hopkins Press; 1965. p. 303–29.
Tosa Y, Chuma I. Classification and parasitic specialization of blast fungi. J Gen Plant Pathol. 2014;80:202–9.
Tufan HA, McGrann GR, Magusin A, Morel JB, Miché L, Boyd LA. Wheat blast: histopathology and transcriptome reprogramming in response to adapted and nonadapted *Magnaporthe* isolates. New Phytol. 2009;184(2):473–84. https://doi.org/10.1111/j.1469-8137.2009.02970.x.
Valent B, Chumley FG. Avirulence genes and mechanisms of genetic instability in the rice blast fungus. In: Zeigler R, Leong S, Teng P, editors. The rice blast disease. Cambridge: International Rice Research Institute, Los Banos and Commonwealth Agricultural Bureaux; 1994. p. 111–4.
Van der Biezen EA, Jones JDG. Plant disease-resistance proteins and the gene-for-gene concept. Trends Plant Sci. 1998;23:454–6.
Xue M, Yang J, Li Z, Hu S, Yao N, Dean RA, et al. Comparative analysis of the genomes of two field isolates of the rice blast fungus *Magnaporthe oryzae*. PLoS Genet. 2012;8(8):e1002869. https://doi.org/10.1371/journal.pgen.1002869.
Yamasaki Y, Kiyosawa S. Studies on inheritance of resistance of rice varieties to blast. 1. Inheritance of Japanese varieties to several strains of the fungus. Bull Natl Inst Agric Sci D. 1966;14:39–69.
Yoshida K, Saitoh H, Fujisawa S, et al. Association genetics reveals three novel avirulence genes from the rice blast fungal pathogen *Magnaporthe oryzae*. Plant Cell. 2009;21(5):1573–91. https://doi.org/10.1105/tpc.109.066324.
Yu ZH, Mackill DJ, Bonman JM, Tanksley SD. Tagging genes for blast resistance in rice via linkage to RFLP markers. Theor Appl Genet. 1996;81:471–6.
Zellerhoff N, Jarosch B, Groenewald JZ, Crous PW, Schaffrath U. Non-host resistance of barley is successfully manifested against *Magnaporthe grisea* and a closely related *Pennisetum*-infecting lineage but is overcome by *Magnaporthe oryzae*. Mol Plant-Microbe Interact. 2006;19:1012–22.
Klaubauf S, Tharreau D, Fournier E, Groenewald JZ, Crous PW, de Vries RP, Marc-Henri L. Resolving the polyphyletic nature of Pyricularia (Pyriculariaceae). Stud Mycol 2014;79:85–120.
Urashima AS, Igarashi S, Kato H. Host range, mating type and fertility of Pyricularia grisea from wheat in Brazil. Plant Dis. 1993;77:1211–6.
Koide Y, Telebanco-Yanoria MJ, Fukuta Y, Kobayashi N. Detection of novel blast resistance genes, *Pi58(t)* and *Pi59(t)*, in a Myanmar rice landrace based on a standard differential system. Mol Breeding. 2013;32:241–52.
Couch BC, Kohn LM. A multilocus gene genealogy concordant with host preference indicates segregation of a new species, Magnaporthe oryzae, from M. grisea. Mycologia. 2002;94:683–693.
Su J, Wang W, Han J, Chen S, Wang C, Zeng L, Feng A, Yang J, Zhou B, Zhu X. Functional divergence of duplicated genes results in a novel blast resistance gene Pi50 at the Pi2/9 locus. Theor Appl Genet. 2015;128(11):2213–25.
Zheng W, Wang Y, Wang L, Ma Z, Zhao J, Wang P, Zhang L, Liu Z, Lu X. Genetic mapping and molecular marker development for Pi65(t), a novel broadspectrum resistance gene to rice blast using next-generation sequencing. Theor Appl Genet. 2016;129(5):1035–44.
Zhu D, Kang H, Li Z, Liu M, Zhu X, Wang Y, Wang D, Wang Z, Liu W, Wang GL. A genome-wide association study of field resistance to *Magnaporthe oryzae* in rice. Rice. 2016;9:44. https://doi.org/10.1186/s12284-016-0116-3.
Cruz CD, Valent B. Wheat blast disease: danger on the move. Trop plant pathol. 2017;42:210–22.
McHale L, Tan X, Koehl P, Michelmore RW. Plant NBS-LRR proteins: adaptable guards. Genome Biol. 2006;7:212. https://doi.org/10.1186/gb-2006-7-4-212.

Chapter 13
Pearl Millet Blast Resistance: Current Status and Recent Advancements in Genomic Selection and Genome Editing Approaches

B. M. Anil Kumar, Rajashekara Hosahatti, C. Tara Satyavathi, Ganesan Prakash, Rajan Sharma, R. Narasimhulu, and S. Chandra Nayaka

Contents

B. M. Anil Kumar · S. Chandra Nayaka (✉)
Pearl Millet Pathology Laboratory, DOS in Biotechnology, ICAR-AICRP, Mysore Centre, University of Mysore, Mysore, India

R. Hosahatti
ICAR-Vivekananda Parvatiya Krishi Anusandhan Sansthan, Almora, Uttarakhand, India

C. T. Satyavathi
ICAR-AICRP on Pearl Millet, Jodhpur, Rajasthan, India

G. Prakash
Division of Plant Pathology, ICAR-Indian Agricultural Research Institute, New Delhi, India

R. Sharma
International Crops Research Institute for the Semi-Arid Tropics (ICRISAT), Patancheru, Hyderabad, Telangana, India

R. Narasimhulu
Acharya NG Ranga Agricultural University Agricultural Research Station, Ananthapuram, Andhra Pradesh, India

S. C. Nayaka et al. (eds.), *Blast Disease of Cereal Crops*, Fungal Biology,
https://doi.org/10.1007/978-3-030-60585-8_13

13.1 Introduction

Pearl millet is a hardy C_4 crop grown in the arid and semiarid tropics mainly in rainfed conditions where other cereals are likely to fail to produce economic yields due to drought and heat stresses. Adaptive evolution, a form of natural selection, shaped the crop to grow and yield satisfactorily with limited moisture supply or under periodic water deficits in the soil (Serba and Yadav 2016). *Pyricularia* leaf spot, also known as blast disease, is particularly important in pearl millet forage cultivars. It is an important disease in the southern United States and more recently it has emerged as a serious disease of dual-purpose (grain and fodder) pearl millet hybrids in India (Lukose et al. 2007; Anonymous 2009). In India, for the first time the disease was recorded from Kanpur, Uttar Pradesh (Mehta et al. 1953), and it was considered as a minor disease for a long time since its inception.

The pathogen infects several cereal crops, including rice, wheat, pearl millet, finger millet, and foxtail millet, and several grasses. The pathogen is highly variable, but highly specialized in its host range. Thus, *M. grisea* strains isolated from rice or any other hosts do not infect pearl millet and vice versa. The rice blast pathosystem has been extensively studied for several decades. In disease-conducive environments, the life span of many disease-resistant rice cultivars has been known to be ephemeral (Srinivasachary et al. 2002). Most of the deployed resistance genes in rice crop break down in a few years because of their race specificity, high rate of mutation, and rapid change in pathogenicity of the blast pathogen (Suh et al. 2009). Pathogenic variation in *M. grisea* populations adapted to rice, finger millet, foxtail millet, wheat, and several weed hosts has been reported by Takan et al. (2012). Various potential mechanisms, including sexual recombination, heterokaryosis, parasexual recombination, and aneuploidy, have been proposed to explain frequent race changes in *M. grisea* (Kang and Lee 2000). This implies that pathogenic variability might exist in the pearl millet-infecting strains of *M. grisea*. Therefore, for the management of disease through host plant resistance, it is important to know pathogen variation in populations of *M. grisea* infecting pearl millet and to identify potential resistance sources.

Disease resistance in plants occurs as a hypersensitive response (HR) at the site of infection by pathogen. This specific event is initiated in response to recognition of pathogen-associated molecular pattern (PAMP) and subsequent PAMP-triggered immunity (PTI) and effector-triggered immunity (ETI) (Boller and He 2009). Both PTI and ETI mechanisms are closely associated with reactive oxygen species (ROS) production and disease resistance that involves production of distinct biphasic ROS as one of its pivotal plant immune mechanisms. This unique oxidative burst is strongly dependent on the resistant cultivars because a monophasic ROS burst is a hallmark of the susceptible cultivars. However, the cause of the differential expression of ROS burst remains still unknown. A second phase of the innate immune system in plants, known as effector-triggered immunity (ETI), was discovered to be mediated by intracellular receptor molecules containing nucleotide-binding (NB) and leucine-rich repeat (LRR) domains that specifically recognize effector proteins produced by the pathogen. The binding of receptors and effectors results in the activation of defense programs and often leads to localized cell death (Chen and Ronald 2011). In fact, ETI reactivation of defense responses is assumed to be mediated by the activity of cytoplasmic proteins and transcriptional reprogramming to make the first immune responses involving signaling from the cytoplasm to the nucleus (Wirthmueller et al. 2007).

Effectors are mainly involved in the colonization of pathogen or interrupt the activation of host cell defenses. On the contrary, rice plants have consequently showed immunity that is based on the sensitivity of effectors to host resistance proteins. *M. grisea* produces effector proteins to influence plant immunity, leading to the penetration of infection process. Avr proteins are a special group of effectors encoded by avirulence genes. These proteins can be distinguished by relative R proteins and this leads to the race-specific recognition (Wawra et al. 2013). Pathogen AVR effectors have been genetically proven to be essential components in plant immune responses (Flor 1971). However, the mechanism by which AVR effectors and R proteins are associated with these responses remains unclear. Initially, the ligand-receptor model (Gabriel and Rolfe 1990) was widely supported, but the lack of physical interactions between a number of R/AVR pairs has resulted in the generation of alternative guard and decoy hypothesis (van der Hoorn and Kamoun 2008). One of the interesting results have been reported recently for AVR-Pii and OsExo70-F3 interaction in which OsExo70-F3 physically interacts with AVR-Pii and is specifically involved in Pii-dependent resistance suggesting OsExo70-F3 as a helper in Pii/AVR-Pii interactions (Fujisaki et al. 2015). More recently, rice-resistant protein pair RGA4/RGA5 was shown to be required for recognition of *M. oryzae* effectors AVR-Pia and AVR1-CO39 revealing a mode of AVR protein recognition through direct binding to a novel, non-LRR interaction domain (Cesari et al. 2013, 2014). *M. oryzae* has evolved effector proteins such as Slp1 and AVR-Piz-t that can suppress rice immune responses, whereas the rice crop has developed an innate immune system that can recognize the presence of *M. oryzae* and initiate PTI and ETI defense responses (Liu et al. 2013; Singh et al. 2016). A model of AVR-Pii and Os-NADP-ME interactions near the BIC after *M. oryzae* infection was generated and it was also reported that the AVR-Pii-Os-NADP-ME2-3 pair provides key evidence on how AVR effectors reprogram host metabolism to establish compatibility and why the ROS issue has continued to be treated as a main defense mechanism in the innate immunity of most plants.

Insights into the function of secreted fungal effectors in the reprogramming of host defense and metabolism are gradually emerging. Among the plethora of secreted proteins encoded in the genomes of phytopathogenic fungi, only a few fungal effector targets have been identified to date, with less than a handful shown to impair host metabolic functions directly (Tanaka et al. 2014). Although eight AVR genes of *M. oryzae* have been isolated over the last several decades (Zhang et al. 2015), their functions as virulence factors have not been functionally addressed, except for evidences found in AVR-Pizt and AVR-Pii (Fujisaki et al. 2015; Park et al. 2012).

In recent years, more progress has been made in investigating the molecular mechanisms of innate immunity responses in rice and other cereal crops against *M. grisea*, as well as in identifying R genes, identification-triggered early signaling, signaling pathways in rice, and role of these signaling pathways in activating defense responses (Saitoh et al. 2011; Seo et al. 2011). However, a complete understanding of the molecular network regulating defense responses against pathogens in rice and millets is still obscured. For instance, despite the number of R genes present in rice, little information is available about the essential signaling needed to initiate effector-triggered resistance against *M. grisea*.

13.2 Plant Impedance and Genome Editing Tools

The traditional genetic breeding methods depend mainly on the existence of broad genetic variation in elite primary gene pool of particular crop species and its closely related crop species to introduce new resistant traits into local gene pool, which is purely naturally induced spontaneous mutation (Jung et al. 2018). The first artificially induced transgenesis was experimentally reported in bacteria by Avery et al. (1944). But for the first time gene modification or genome editing tools with respect to plants started during early 1980s. The work of Schilperoort et al. (1967) on *Agrobacterium tumefaciens* has become the key step in understanding the transfer of genes through the Ti plasmid (200 kb) of virulent strain and its integration into nuclear chromosomes (Zaenen et al. 1974; Van Larebeke et al. 1974; Chilton et al. 1980). Earlier to this a group of investigators at the Institute of Pasteur in Paris discovered a unique phenomenon of deviation from the Mendel's laws of genetic inheritance when they were studying the crossing in the yeast: they discovered that a particular mitochondrial allele, known as omega (ω), was unidirectionally inherited, which was later determined to be the gene for the large ribosomal RNA subunit which is further known to be type 1 intron; till then introns were regarded as junk DNA; after a few days, it became evident that this omega is encoded by separate gene producing an endonuclease which recognizes and introduces the double-strand break (DSB) over the LSU rRNA genes lacking the intron; these DNA nicks are repaired by DNA repair mechanism. This became the first basic foundation of homing meganucleases to use them in genetic engineering (Bolotin et al. 1971; Bos et al. 1978; Faye et al. 1979; Jacquier and Dujon 1985). But there was still a question as to how to introduce nick in the double-stranded DNA with unique locus of choice interest, which was addressed by Snyter and Brooks (1988) and his colleagues through the experimental characterization of best rare cutting endonuclease *Not1* enzyme with recognition site of 8 bp length and the I-Sce I endonuclease (omega of LSUrRNA) study pioneered the accuracy and precisions of editing of plant genomes (Puchta et al. 1993).

As per the records reported till now, for the first time plant genome editing was reported from *Agrobacterium* through chemical mutagenesis with the help of ethyl methane sulfonate (EMS) and ionizing radiations which created genome modifications randomly. The second phase is associated with the discoveries of homing and meganuclease enzymes during the 1980s and 1990s as mentioned earlier, which are engineered to provide efficient tools for targeted editing in genome. During 2006–2012, a few crop plants were successfully and precisely modified using zinc-finger nucleases. A third phase (2009–2011) of improvements in genome editing led to a dramatic decrease in off-target events with the TALEN technology. In the year 2013, major breakthrough came with the development of CRISPR-Cas9 technology which has high efficiency and technical essence of use is quite impressive; this technology can be employed with the use of in-house kits or commercially available kits. The main requirements include careful selection of the location of the DNA double-strand breaks to be induced and then ordering of sequence of an oligonucleotide usually a primer (Francis 2016). Genome engineering is currently done by employing three different tools as shown in Fig. 13.1.

As it is known gene editing tools have the ability to implement DBS's double-strand breaks in the genome, which further requires repair of these breaks for a successful knock-in and knockout events. Subsequently, the damaged DNA is repaired mainly by two molecular repair mechanisms: homology-directed repair (HDR), where the broken DNA is repaired using a homologous DNA sequence as template, or nonhomologous end joining (NHEJ) where the broken ends are rejoined to each other at nonhomologous DNA sequence and NHEJ is mainly responsible for repair of one or two nucleotide breaks in the DNA leading to gene disruption whereas HDR is responsible for gene deletion, replacement, and gene insertion (as shown in Fig. 13.2). HDR works by precise copying of gene with the help and template at specific site, repairs the homologous DNA breaks, and is helpful for gene expression studies whereas NHEJ repair mechanism only repairs by insertion or

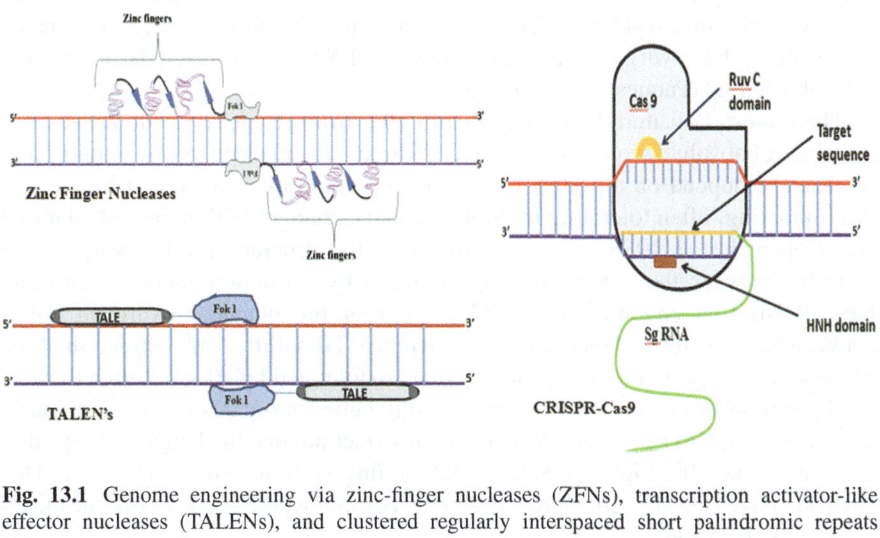

Fig. 13.1 Genome engineering via zinc-finger nucleases (ZFNs), transcription activator-like effector nucleases (TALENs), and clustered regularly interspaced short palindromic repeats (CRISPR) and Cas 9

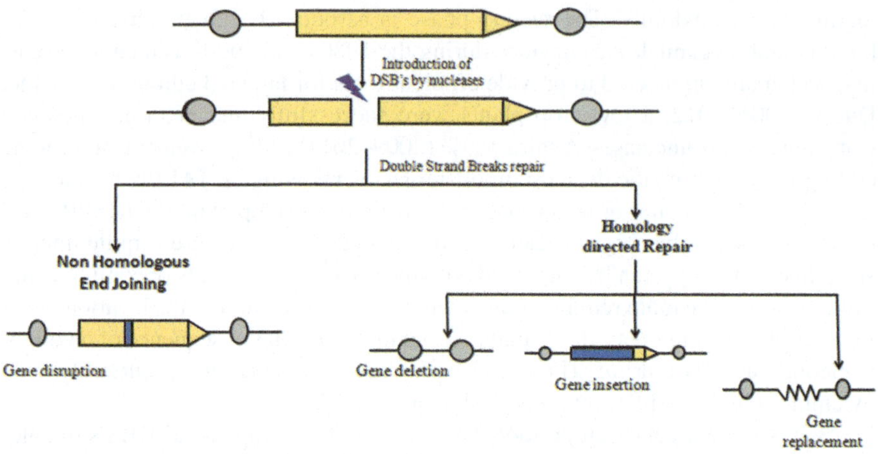

Fig. 13.2 DNA repair mechanism induced by double-stranded breaks leading to gene disruption via NHEJ pathway and gene deletion, insertion, and replacement by HDR mechanisms

deletion of one or two nucleotides where it is helpful in the SNP, knock-in, and knockout studies (Maresca et al. 2013; Lee et al. 2015).

13.3 Meganucleases or Homing Endonucleases (HEs)

HEs are small proteins (<300 amino acids) found in bacteria, archaea, and unicellular eukaryotes. A distinguishing characteristic of HEs is that they recognize relatively long sequences (14–40 bp) compared to other site-specific endonucleases such as restriction enzymes (4–8 bp). These lengthy recognition sites, and the name of the first such known enzyme, ω (also known as I-SceI), have given rise to the term "meganuclease" (Paques and Duchateau 2007).

The important feature that differentiates HEs from restriction endonucleases is their lack of absolute sequence specificity. Whereas restriction enzyme binding and/or cleavage depend on a perfect match to the recognition sequence, HEs are less discriminating, often tolerating multiple sequence changes within their recognition site (Colleaux et al. 1988). This is significant at the structural level making a great diversity between the number of contacts made by restriction endonucleases and HEs. Restriction endonucleases exploit most of the potential hydrogen bonds between the proteins and their target sites whereas HEs utilize only a fraction of the possible hydrogen bonds. The positions that are tolerated by HEs are often those at third positions of codons, which vary naturally between organisms. Such tolerance allows homing into new sites. Despite the imperfect fidelity, the lengthy recognition sites can make HEs highly specific, often cutting large genomes only once. This attribute makes the HEs amenable to genome editing, where spurious off-site cleavages are detrimental.

HEs have been historically categorized by small conserved amino acid motifs. At least five such families have been identified: LAGLIDADG, GIY-YIG, HNH, His-Cys Box, and PD – (D/E) × K, which are related to ED × HD enzymes and are considered by some as a separate family. At a structural level, the HNH and His-Cys Box share a common fold (designated ββα-metal) as same as the PD – (D/E) × K and ED × HD enzymes. The catalytic and DNA recognition strategies for each of the families vary and lend themselves to different degrees to engineering for a variety of applications among which LAGLIDADG homing endonucleases are well known (Belfort and Bonocora 2014). Another focus was directed to create hybrids which provide platform for increasing the specificity, employed for GIY-YIG and LAGLIDADG groups of enzymes. The idea of hybrids helped to produce catalytically nonactive with interest of recognition site enzymes coupling with catalytically active with undefined recognition site. The chimeric construct catalytically inactive LHE I-SceI fused to the restriction enzyme PvuII as the cleavage module where the two subunits are joined by linker (Fonfara et al. 2012). TALE-PvuII chimeric protein was also developed after the discovery of TALENS to increase the efficiency of inducing the breaks at a particular site (Yanik et al. 2013).

13.4 Zinc-Finger Endonucleases (Nucleases), ZFNs

ZFNs are artificially developed restriction enzymes that have been successfully used for a few years as a genome editing tool. As it is known that two most desired qualities of genome editing comprise (1) an endonuclease that must be able to recognize specific long target sequences and (2) adequate flexibility of re-targeting to desired sequences defined by use. The ZFN structure meets the above properties better than any other meganucleases. The ZFN carries the DNA-binding domain zinc-finger proteins (ZFPs) derived from eukaryotic transcription factors and the Fok I cleavage domain derived from *Flavobacterium okeanokoites* (Bitinaite et al. 1998; De Souza 2011). The zinc-finger nuclease is composed of 30 amino acids forming two antiparallel β-sheets opposite an alpha-helix. Each ZFN cleavage domain is linked to a range of three–six zinc fingers designed to specially recognize the target sequence flanking the cleavage site. The most commonly used zinc fingers have three DNA-binding domains, and nine base pairs. The zinc-finger proteins are used in pairs giving 18-base specificity of each bind to a single codon (i.e., three nucleotide codons). ZFN can become an optional tool for researchers, but due to the high cost and design complications, presently ZFNs have been successfully used for HDR-mediated gene KI and NHEJ-mediated KO approaches to many targets in prokaryotes as well as eukaryote gene editing experiments (Carroll 2011). Even though with so much advances in engineered systems, they also have some drawbacks like all possible nucleotide sequences of the genome cannot be targeted and its specificity to the sequence cannot be precisely defined as it may be influenced by other neighboring protein domains (Maeder et al. 2008).

Despite their complicated construction and off-targeting issues, ZFNs have already been used to make gene modifications in Arabidopsis (Osakabe et al. 2010; Petolino et al. 2010; Zhang et al. 2010; Even-Faitelson et al. 2011; Qi et al. 2013), tobacco (*Nicotiana tabacum*) (Wright et al. 2005; Townsend et al. 2009), and also in crops including maize (*Zea mays*), soybean (*Glycine max*), and canola (*Brassica napus*) (Shukla et al. 2009; Curtin et al. 2011; Gupta et al. 2012; Ainley et al. 2013). Resistance to bialaphos in maize (Shukla et al. 2009), resistance to herbicides in tobacco (Townsend et al. 2009), and ABA-insensitive phenotype in Arabidopsis (Osakabe et al. 2010). However, ZFNs lack specificity which results in the generation of undesired mutations, off targets, and some chromosomal abbreviations (Pattanayak et al. 2011; Radecke et al. 2010).

13.5 Transcription Activator-Like Effector Nucleases (TALENS)

After several years, Schornack et al. (2006) reported some proteins from *Xanthomonas* bacterial species that are capable of hijacking the plant immune response in rice. These proteins are 30–35-residue-long sequences with most positions occupied by conserved sequences and a series of tandem repeats, of which four conserved positions are with limited variability [two amino acids in positions 12 and 13 (called "diresidues") vary and specify the binding to A, C, G, and T as shown in Fig. 13.2] with DNA-binding domain and an effector domain. Proteins with similar properties were reported and were termed as transcription activator-like effector (TALE) (Moscou and Bogdanove 2009; Boch et al. 2009). The TALE proteins are engineered with a FokI nuclease domain which creates the DSB (Mussolino and Cathomen 2012). TALENs possess longer recognition sites which help in increasing the specificity and then ZFNs are highly efficient with less prone to mutations and cause very minimum deleterious effects (Puchta and Fauser 2014; Petersen and Niemann 2015). The plant pathogen TALE's N-terminal segment (NTS) harbors protein secretion signal peptides while their C-terminal segment (CTS) contains nuclear localization signal peptides and a transcription activator domain which play a very important role in cleavage efficiency by impairing the catalytic domain scaffold when fused with the FokI, so the scaffold optimization plays an important role during engineering the TALENs for developing efficient genome editing tool with precise on targets and DSB of our interest (White et al. 2009).

Just like ZFNs, TALEN construction also requires verification in vitro to have the knowledge of efficiency of site-specific cleavage instead of directly employing into the experiments because of the modular nature of the various segments (conserved sequences, NTS and CTS) in TALEN; the probability of rationally designing a TALEN that will be specific for a particular sequence appears to be greater than that for ZFNs; however, it is not possible to rule out off-target cleavage in an experiment model without extensive testing and confirmation.

Rice bacterial blight is caused by *Xanthomonas oryzae* pv. *oryzae* (Xoo) and its pathogenesis is controlled by the interaction TALE of *Xanthomonas oryzae* and host target S genes. TALENs were developed with respect to S gene effector-binding element site (EBE) and induced the disruption of EBE which resulted in lack of interaction between *Xanthomonas* TALE protein and S genes resulting in subsequent establishment of blight resistance rice line (Li et al. 2012). The first TALEN-modified rice was made by mutating the OsSWEET14 promoter to generate heritable and effective disease-resistant lines. Powdery mildew pathogens cause yield loss up to 50% worldwide (Cao et al. 2011) and Wang et al. (2014) attempted to obtain a resistant variety of wheat by targeting the *mlo* (mildew locus O) gene coding for proteins repressing defense against the disease. A pair of TALENs was created that targeted a region in exon 2 of all three homolog genes, creating mainly small deletions. Homozygous mutation of all three homolog genes was necessary to have an effective and heritable powdery mildew resistance.

13.6 Clustered Regularly Interspaced Short Palindromic Repeats (CRISPR) and Associated Systems

Similar to eukaryotic mammalian immune system even bacteria and most of the archaea harbor a special type of immune response against various infections. Bacteria and archaea bacteria of known being are invaded by phages; bacteria fend this invasion of foreign bodies by employing clustered regularly interspaced short palindromic repeats (CRISPR) and CRISPR-associated proteins, which is similar to adaptive immunity of eukaryotic immune system; hence it is also referred to as adaptive immunity of bacteria and archaea bacteria (Jansen et al. 2002; Sorek et al. 2008). In eukaryotes the antigen and presentation are done by classical antigen-presenting cells, where a piece of fragment of that foreign body is being sliced and represented to recognize and defend the infection similarly in bacteria; as soon as it encounters the phage infection the bacterium with CRISPR acquires a piece of its DNA and tailors close to promoter sequence of CRISPR array. CRISPR array transcribes and generates m-RNA (referred to as crRNA) which serves as a guiding molecule for Cas9 proteins (helicase and nuclease proteins) to the target phage genome, cleaves the special nucleotide sequence (3–5 base pair) found upstream to protospacer adjacent motif (PAM) sequence in future invasion, and provides immunity to bacterial cell (Jinek et al. 2012). Even though CRISPR repeats were initially discovered in the 1980s in *E. coli* its function remained ununderstood (Ishino et al. 1987) but experimentally demonstrated resistance of *S. thermophilus* acquired against a bacteriophage by integrating a virus genomic fragment into its CRISPR locus (Barrangou et al. 2007).

Cas9 protein was characterized and known to have two catalytically active domains, namely N-terminal HNH domain with nuclease domain with histidine and asparagine residues in the middle region and RuvC domain with nuclease activity towards C-terminal end (Sapranauskas et al. 2011).

The CRISPR element consists of a complex of noncoding RNAs and Cas proteins (CRISPR associated), which have nuclease and helicase activity. In contrast to the chimeric TALEN proteins, recognition by the CRISPR/Cas system is carried out via the complementary interaction between noncoding RNA and target-site DNA, whereas TALEN tool is protein-directed gene editing. Simple construct of CRISPR plasmid contains elements necessary for CRISPR/Cas9 activity: U3 or U6 RNA polymerase III promoters, Cas9 nuclease/nickase, CRISPR mRNA, and tracrRNA. The crRNAs consist of approximately 40 bp sequences in association with tracrRNA that activates and guides Cas9 nuclease. Biolistic and agrobacterium-mediated transformation is being employed to transfer the sgRNA and Cas9 proteins into the desired cell when it is used for plant genome editing. The target sequence of choice of pathogen is inserted into this position by introducing DSB in the plasmid by employing site-specific endonucleases; once the cell encounters the invasion by respective pathogen this CRISPR-Cas9 protein gets activated and induces the cleavage in the target genome which is referred to as protospacers (Nemudryi et al. 2014; Barrangou et al. 2007). The advantages of CRISPR-Cas gene editing technology over other tools are that protein engineering is not required like ZFNs or TALENs, modification like DNA methylation can be done which cannot be done using ZFNs or TALENs, multiple targeting (multiplexing) can be done in single experiment, and even any number of gRNAs can be designed for a single target which makes it special (Ding et al. 2013; Doench et al. 2014; Mao et al. 2013).

One of the most powerful applications of CRISPR-Cas technology in plants will be to study the specific gene function involved in the metabolic or developmental pathways for knockout. For example, two key proteins in photosynthesis, highly homologous magnesium chelatase subunit I (CHLI) genes, CHLI1 and CHLI2, have been simultaneously targeted for elimination and have resulted in the production of albino plants as a useful tool for understanding the role of these enzymes in chlorophyll biosynthesis (Mao et al. 2013). Understanding the role of a particular gene in plant development like the auxin-binding protein (ABP1) is not required for either auxin signaling or Arabidopsis development (Gao et al. 2015). These two examples showed the precise elimination of one or more specific genes of character of interest to study the effects of its gene product role in the regulation of plant growth and development. Similarly multiplexing can be done because gRNA is small in size so that many genes can be targeted simultaneously. Golden gateway and Gibson assembly are the two important techniques which are employed to make multiplex targeting. Gibson assembly works based on primer overhangs whereas golden gateway of cloning works by employing restriction endonucleases which helps to target 5–8 genes. The other way of modifying plant growth and development and maximum yield of food grains for commercial benefit is to manipulate the specific gene timing and level of its expression by employing dead Cas9 (lacks the nuclease activity) coupling it with gene activator and gene repressor proteins, which guides the polymerase to transcribe active and continuously or repress the expression continuously. The important issue after transformation is screening of positive clones with different techniques that have been developed, like annealing at critical temperature-polymerase chain reaction (ACT-PCR) (Hua et al. 2017), high-resolution melting analysis (HRMA) (Thomas et al. 2014), polyacrylamide gel electrophoresis (PAGE)-mediated genotyping (Zhu et al. 2014), T7 endonuclease I (T7EI) approach (Vouillot et al. 2015), and restriction enzyme site loss technique. After the successful editing of genome of Arabidopsis (Feng et al. 2013; Li et al. 2013) and rice (Miao et al. 2013; Shan et al. 2013) using CRISPR/Cas9 system has been reported, many independent applications of this genome editing tool in various crops have shown that CRISPR/Cas9 system is broadly acceptable and effective for crop improvement like generating herbicide resistance, disease resistance, enhancing of yield levels, and stress tolerance.

13.7 Development of Blast Resistance by Genome Editing

Potato crop is harvested and stored in cold rooms to avoid sprouting; due to storage under cold conditions the starch of potato degrades into glucose and fructose. These reduced sugars, when subjected to deep frying in oil, i.e., while making chips, French fries, etc., turn into a carcinogen called acrylamide. Chawla et al. (2012) used RNAi to overcome this problem and developed a new variety of potato called “innate potato”; similarly another group of scientists used a gene of enzyme called *vascular invertase* which plays a very important role in reducing the sucrose into glucose and fructose by using TALENs; it was successful in providing potatoes with very less starch and helps in brown-colored acrylamide formation (Clasen et al. 2016). Similarly, when apple cut was exposed to air it turned into brown color, because of reduction in polyphenols by polyphenol oxidase enzyme (PPO gene). RNAi was used to address this reduction by blocking the homologous *PPO* gene expression (Waltz et al. 2015). Herbicide-resistant crops were also developed by inducing the mutation in acetolactate synthase (ALS) gene which serves as a target for imidazolinone herbicides (Townsend et al. 2009) by employing the ZFNs. Citrus canker caused by *Xanthomonas citri* is a serious disease in many citrus cultivars, leading to huge economic losses around the world. Pathogen injects its effector (TALEPthA4) protein which has target of CsLOB1 gene EBE motifs resulting in active transcription of canker susceptibility products. It is known to possess two genes CsLOB1 and two genes where both are capable of functioning in the absence of other (Jia et al. 2016, 2017) induced mutation in both genes and generated canker-resistant citrus plants, which bind to EBE motifs and transcriptionally activate the downstream target canker susceptibility lateral organ boundary 1 (CsLOB1) gene in the host, leading to disease susceptibility using CRISPR Cas system. Rice blast is the most devastating in all the rice-growing countries, with yield loss of 10–30% (Dean et al. 2012). Therefore, creating resistance to *M. oryzae* is one of the most effective approaches to mitigate the disease. The plant ethylene response factors

(ERF) are the main factors involved in multiple stress tolerance (Müller and Munné-Bosch 2015). The work of knockdown expression of rice ERF92 by employing RNAi enhances resistance to *M. oryzae* (Liu et al. 2012). CRISPR/Cas9 is used to mutate the OsERF922 gene through agrobacterium-mediated transformation of rice calli (Kuiku 131) where one sgRNA was designed for targeting the first exon of the gene, 21 mutant lines are obtained, and these mutant lines are evaluated for different agronomic parameters, but these mutant lines showed enhanced resistance to *M. oryzae* infection compared to the wild type. The lesion length in the mutant lines is about 66% smaller than that in the wild type (Wang et al. 2016). Ma et al. (2018) developed 273 ossec3a mutants that are more tolerant to *M. oryzae* infection by using CRISPR/Cas9. In the same way even pearl millet is highly affected by blast disease caused by *M. grisea* which can be managed by developing the resistant lines by employing genetic engineering tools. CRISPR-Cas9 tool can be used to target the genes responsible for the susceptibility to blast disease which are commonly termed as S genes and induce the DSBs in S gene by Cas system and direct it towards any of the DNA repair mechanism, resulting in the mutation of S gene which yields blast-resistant pearl millet crop lines (as shown in Fig. 13.3).

13.8 Conclusion and Future Prospects

Pearl millet blast severity has been increasing in recent years due to adoption of high-yielding varieties and use of chemical fungicides is not safe to environment. The use of recent technology CRISPR/Cas9 system which is simple, efficient, and highly specific produces fewer off-target events. It is a promising tool for genome modification in crop improvement program. Mutation in host susceptible gene/s by using CRISPR/Cas9 system has conferred broad-spectrum disease resistance against fungal diseases. This technology is successfully utilized in powdery mildew resistance locus O (*mlo*) which is one of the renowned S genes. It has been mutated via CRISPR/Cas9 and the mutants obtained were more resistant to powdery mildew fungus *Blumeria graminis* f. sp. *tritici* (Bgt) and *Oidium neolycopersici* in wheat and tomato, respectively. This technology is applied in rice blast which is one of the most damaging diseases in the world, which is caused by *Magnaporthe oryzae*. Targeted loss-of-function mutagenesis of negative regulator *Oryza sativa* ethylene response factor 922 (OsERF922) showed enhanced disease resistance against rice blast. Similarly, mutation in coding region of Pi21 gene via CRISPR/Cas9 has also improved the resistance in rice plants against blast fungus. This technology was utilized in rice blast resistance development with the help of the coding region of mitogen-activated protein kinase 5 (OsMAPK5). This technology has lot of potential in the development of blast disease resistance in pearl millet crop with suitable modifications in the genome.

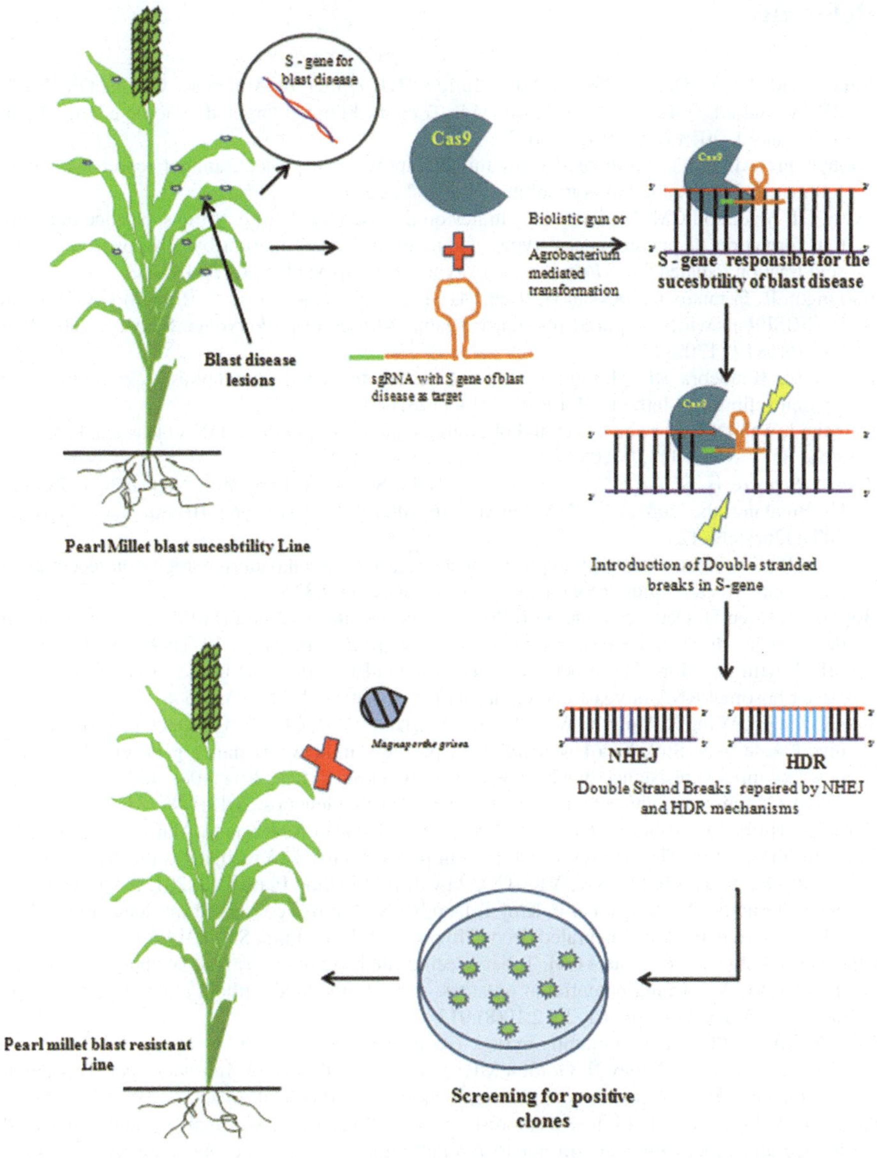

Fig. 13.3 Typical overview of generating blast disease-resistant pearl millet crop lines using CRISPR-Cas 9 system

References

Ainley WM, Sastry-Dent L, Welter ME, Murray MG, Zeitler B, Amora R, Corbin DR, Miles RR, Arnold NL, Strange TL, Simpson MA. Trait stacking via targeted genome editing. Plant Biotechnol J. 2013 Dec;11(9):1126–34.

Anonymous. All India coordinated pearl millet improvement project, annual report. Mandore, Jodhpur: Indian Council of Agricultural Research; 2009.

Avery OT, MacLeod CM, McCarty M. Studies on the chemical nature of the substance inducing transformation of pneumococcal types: induction of transformation by a desoxyribonucleic acid fraction isolated from *Pneumococcus* type III. J Exp Med. 1944;79:137–58.

Barrangou R, Fremaux C, Deveau H, Richards M, Boyaval P, Moineau S, Romero DA, Horvath P. CRISPR provides acquired resistance against viruses in prokaryotes. Science. 2007 Mar 23;315(5819):1709–12.

Belfort M, Bonocora RP. Homing endonucleases: from genetic anomalies to programmable genomic clippers. Methods Mol Biol. 2014;1123:1–26.

Bitinaite J, Wah D, Aggarwal AK, et al. FokI dimerization is required for DNA cleavage. Proc Natl Acad Sci U S A. 1998;95:10570–5.

Boch J, Scholze H, Schornack S, Landgraf A, Hahn S, Kay S, Lahaye T, Nickstadt A, Bonnas U. Breaking the code of DNA binding specificity of TAL-Type III effectors. Science. 2009;326:1509–12.

Boller T, He SY. Innate immunity in plants: an arms race between pattern recognition receptors in plants and effectors in microbial pathogens. Science. 2009;324:742–4.

Bolotin M, Coen D, Deutsch J, Dujon B, Netter P, Petrochilo E, Slonimski PP. La recombinaison des mitochondries chez *Saccharomyces cerevisiae*. Bull Inst Pasteur. 1971;69:215–39.

Bos JL, Heyting C, Borst P, Arnberg AC, van Bruggen EFJ. An insert in the single gene for the large ribosomal RNA in yeast mitochondrial DNA. Nature. 1978;275:336–8.

Cao A, Xing L, Wang X, Yang X, Wang W, Sun Y, Qian C, Ni J, Chen Y, Liu D, et al. Serine/threonine kinase gene Stpk-V, a key member of powdery mildew resistance gene Pm21, confers powdery mildew resistance in wheat. Proc Natl Acad Sci U S A. 2011;108:7727–32.

Carroll D. Genome engineering with zinc-finger nucleases. Genetics. 2011;188:773–82.

Cesari S, Thilliez G, Ribot C, Chalvon V, Michel C, Jauneau A, Rivas S, Alaux L, Kanzaki H, Okuyama Y, et al. The rice resistance protein pair RGA4/RGA5 recognizes the *Magnaporthe oryzae* effectors AVR-Pia and AVR1-CO39 by direct binding. Plant Cell. 2013;25:1463–81.

Cesari S, Bernoux M, Moncuquet P, Kroj T, Dodds PN. A novel conserved mechanism for plant NLR protein pairs: the "integrated decoy" hypothesis. Front Plant Sci. 2014;5:606.

Chawla R, Shakya R, Rommens CM. Tuber-specific silencing of asparagine synthetase-1 reduces the acrylamide-forming potential of potatoes grown in the field without affecting tuber shape and yield. Plant Biotechnol J. 2012;10(8):913–24.

Chen X, Ronald PC. Innate immunity in rice. Trends Plant Sci. 2011;16:451–9.

Chilton MD, Saiki RK, Yadav N, Gordon MP, Quétier F. T-DNA from *Agrobacterium* Ti plasmid is in the nuclear DNA fraction of crown-gall tumor cells. Proc Natl Acad Sci. 1980;77:4060–4.

Clasen BM, Stoddard TJ, Luo S, Demorest ZL, Li J, Cedrone F, Coffman A. Improving cold storage and processing traits in potato through targeted gene knockout. Plant Biotechnol J. 2016;14(1):169–76.

Colleaux L, D'Auriol L, Galibert F, Dujon B. Recognition and cleavage site of the intron-encoded *omega* transposase. Proc Natl Acad Sci U S A. 1988;85:6022–6.

Curtin SJ, Zhang F, Sander JD, Haun WJ, Starker C, Baltes NJ, Reyon D, Dahlborg EJ, Goodwin MJ, Coffman AP, Dobbs D. Targeted mutagenesis of duplicated genes in soybean with zinc-finger nucleases. Plant Physiol. 2011 Jun 1;156(2):466–73.

De Souza N. Primer: genome editing with engineered nucleases. Nat Methods. 2011;9:27.

Dean R, Van Kan JAL, Pretorius ZA, Hammond-Kosack KE, Di Pietro A, Spanu PD, Rudd JJ, Dickman M, Kahmann R, Ellis J, Foster GD. The top 10 fungal pathogens in molecular plant pathology. Mol Plant Pathol. 2012;13(4):414–30.

Ding Q, Lee YK, Schaefer EK. A TALEN genome editing system for generating human stem cell-based disease models. Cell Stem Cell. 2013;12:238–51.
Doench JG, Hartenian E, Graham DB. Rational design of highly active sgRNAs for CRISPR-Cas9-mediated gene inactivation. Nat Biotechnol. 2014;32:1262–7.
Even-Faitelson L, Samach A, Melamed-Bessudo C, Avivi-Ragolsky N, Levy AA. Localized egg-cell expression of effector proteins for targeted modification of the Arabidopsis genome. Plant J. 2011 Dec;68(5):929–37.
Faye G, Dennebouy N, Kujawa C, Jacq C. Inserted sequence in the mitochondrial 23S ribosomal RNA gene of the yeast *Saccharomyces cerevisiae*. Mol Gen Genet. 1979;168:101–9.
Feng Z, Zhang B, Ding W, Liu X, Yang D, Wei P, Cao F, Zhu S, Zhang F, Mao Y, Zhu J. Efficient genome editing in plants using a CRISPR/Cas system. Cell Res. 2013;23:1229–32.
Flor HH. Current status of the gene-for-gene concept. Annu Rev Phytopathol. 1971;9:275–96.
Fonfara I, Curth U, Pingoud A, Wende W. Creating highly specific nucleases by fusion of active restriction endonucleases and catalytically inactive homing endonucleases. Nucleic Acids Res. 2012;40:847–60.
Francis Q. The CRISPR-Cas9 technology: closer to the ultimate toolkit for targeted genome editing. Plant Sci. 2016 Jan 1;242:65–76.
Fujisaki K, Abe Y, Ito A, Saitoh H, Yoshida K, Kanzaki H, Kanzaki E, Utsushi H, Yamashita T, Kamoun S, Terauchi R. Rice Exo70 interacts with a fungal effector, AVR-Pii, and is required for AVR-Pii-triggered immunity. Plant J. 2015;83:875–87.
Gabriel DW, Rolfe BG. Working models of specific recognition in plant-microbe interactions. Annu Rev Phytopathol. 1990;28:365–91.
Gao Y, Zhang Y, Zhang D, Dai X, Estelle M, Zhao Y. Auxin binding protein 1 (ABP1) is not required for either auxin signaling or *Arabidopsis* development. Proc Natl Acad Sci U S A. 2015;112:2275–80.
Gupta SK, Rai AK, Kanwar SS, Sharma TR. Comparative analysis of zinc finger proteins involved in plant disease resistance. PLoS One. 2012 Aug 15;7(8):e42578.
Hua Y, Wang C, Huang J, Wang K. A simple and efficient method for CRISPR/Cas9-induced mutant screening. J Genet Genomics. 2017;44:207–13.
Ishino Y, Shinagawa H, Makino K, Amemura M, Nakata A. Nucleotide sequence of the iap gene, responsible for alkaline phosphatase isozyme conversion in *Escherichia coli*, and identification of the gene product. J Bacteriol. 1987;169:5429–33.
Jacquier A, Dujon B. An intron-encoded protein is active in a gene conversion process that spreads an intron into a mitochondrial gene. Cell. 1985;41:383–94.
Jansen R, Embden JDAV, Gaastra W, Schouls LM. Identification of genes that are associated with DNA repeats in prokaryotes. Mol Microbiol. 2002;43:1565–75.
Jia H, Orbovic V, Jones JB, Wang N. Modification of the PthA4 effector binding elements in type I CsLOB1 promoter using Cas9/sgRNA to produce transgenic Duncan grape fruit alleviating XccDpthA4:dCsLOB1.3 infection. Plant Biotechnol J. 2016;14:1291–301.
Jia H, Zhang Y, Orbovic V, Xu J, White FF, Jones JB, et al. Genome editing of the disease susceptibility gene CsLOB1in citrus confers resistance to citrus canker. Plant Biotechnol J. 2017;15:817–23.
Jinek M, Chylinski K, Fonfara I, Hauer M, Doudna JA, Charpentier E. A programmable dual-RNA-guided DNA endonuclease in adaptive bacterial immunity. Science. 2012;337:816–21.
Jung C, Capistrano-Gossmann G, Braatz J, Sashidhar N, Melzer S. Recent developments in genome editing and applications in plant breeding. Plant Breed. 2018;137(1):1–9.
Kang S, Lee YH. Population structure and race variation of the rice blast fungus. Plant Pathol J. 2000;16:1–8.
Lee JS, Kallehauge TB, Pedersen LE, et al. Site-specific integration in CHO cells mediated by CRISPR/Cas9 and homology-directed DNA repair pathway. Sci Rep. 2015;5:8572.
Li T, Liu B, Spalding MH, Weeks DP, Yang B. High-efficiency TALEN-based gene editing produces disease-resistant rice. Nat Biotechnol. 2012;30(5):390–2.
Li J, Norville JE, Aach J, McCormack M, Zhang D, Bush J, Church GM, Sheen J. Multiplex and homologous recombination-mediated genome editing in *Arabidopsis* and *Nicotiana benthamiana* using guide RNA and Cas9. Nat Biotechnol. 2013;31:688–91.

Liu D, Chen X, Liu J, Ye J, Guo Z. The rice ERF transcription factor OsERF922 negatively regulates resistance to Magnaporthe oryzae and salt tolerance. J Exp Bot. 2012 Jun 13;63(10):3899–911.

Liu W, Liu J, Ning Y, Ding B, Wang X, Wang Z, Wang GL. Recent progress in understanding PAMP-and effector-triggered immunity against the rice blast fungus Magnaporthe oryzae. Mol Plant. 2013 May 1;6(3):605–20.

Lukose CM, Kadvani DL, Dangaria CJ. Efficacy of fungicides in controlling blast disease of pearl millet. Indian Phytopathol. 2007;60:68–71.

Ma J, Chen J, Wang M, Ren YL, Wang S, Lei CL, Cheng ZJ, Sodmergen. Disruption of OsSEC3A increases the content of salicylic acid and induces plant defense responses in rice. J Exp Bot. 2018;69(5):1051–64.

Maeder ML, Thibodeau-Beganny S, Osiak A, et al. Rapid "open-source" engineering of customized zinc finger nucleases for highly efficient gene modification. Mol Cell. 2008;31:294–301.

Mao Y, Zhang H, Xu N. Application of the CRISPR-Cas system for efficient genome engineering in plants. Mol Plant. 2013;6:2008–11.

Maresca M, Lin VG, Guo N. Obligate ligation-gated recombination (ObLiGaRe): custom-designed critical reviews in biotechnology 11 nuclease-mediated targeted integration through non-homologous end joining. Genome Res. 2013;23:539–46.

Mehta PR, Singh B, Mathur SC. A new leaf spot disease of bajra (*Pennisetum typhoides* Staph and Hubbard) caused by a species of *Pyricularia*. Indian Phytopathol. 1953;5:140–3.

Miao J, Guo D, Zhang J, Huang Q, Qin G, Zhang X, Wan J, Gu H, Qu L. Targeted mutagenesis in rice using CRISPR-Cas system. Cell Res. 2013;23:1233–6.

Moscou MJ, Bogdanove AJ. A simple cipher governs DNA recognition by TAL effectors. Science. 2009;326:1501.

Müller M, Munné-Bosch S. Ethylene response factors: a key regulatory hub in hormone and stress signaling. Plant Physiol. 2015;169(1):32–41.

Mussolino C, Cathomen T. TALE nucleases: tailored genome engineering made easy. Curr Opin Biotechnol. 2012;23:644–50.

Nemudryi AA, Valetdinova KR, Medvedev SP, Zakian SM. TALEN and CRISPR/Cas genome editing systems: tools of discovery. Acta Naturae (англоязычная версия). 2014;6(3) (22)

Osakabe K, Osakabe Y, Toki S. Site-directed mutagenesis in Arabidopsis using custom-designed zinc finger nucleases. Proc Natl Acad Sci. 2010 Jun 29;107(26):12034–9.

Paques F, Duchateau P. Meganucleases and DNA double-strand break-induced recombination: perspectives for gene therapy. Curr Gene Ther. 2007;7:49–66.

Park CH, Chen S, Shirsekar G, Zhou B, Khang CH, Songkumarn P, Afzal AJ, Ning Y, Wang R, Bellizzi M, Valent B, Wang GL. The *Magnaporthe oryzae* effector AvrPiz-t targets the RING E3 ubiquitin ligase APIP6 to suppress pathogen-associated molecular pattern-triggered immunity in rice. Plant Cell. 2012;24:4748–62.

Pattanayak V, Ramirez CL, Joung JK, Liu DR. Revealing off-target cleavage specificities of zinc-finger nucleases by in vitro selection. Nat Methods. 2011;8:765–70.

Petersen B, Niemann H. Advances in genetic modification of farm animals using zinc-finger nucleases (ZFN). Chromosom Res. 2015;23:7–15.

Petolino JF, Worden A, Curlee K, Connell J, Moynahan TL, Larsen C, Russell S. Zinc finger nuclease-mediated transgene deletion. Plant Mol Biol. 2010 Aug 1;73(6):617–28.

Puchta H, Fauser F. Synthetic nucleases for genome engineering in plants: prospects for a bright future. Plant J. 2014;78:727–41.

Puchta H, Dujon B, Hohn B. Homologous recombination in plant cells is enhanced by in vivo double strand breaks into DNA by a specific endonuclease. Nucleic Acids Res. 1993;21:5034–40.

Qi Y, Li X, Zhang Y, Starker CG, Baltes NJ, Zhang F, Sander JD, Reyon D, Joung JK, Voytas DF. Targeted deletion and inversion of tandemly arrayed genes in Arabidopsis thaliana using zinc finger nucleases. G3: Genes, Genomes, Genetics. 2013 Oct 1;3(10):1707–15.

Radecke S, Radecke F, Cathomen T, Schwarz K. Zinc-finger nuclease induced gene repair with oligodeoxynucleotides: wanted and unwanted target locus modifications. Mol Ther. 2010;18:743–53.

Saitoh H, Fujisawa S, Mitsuoka C, Ito A, Hirabuchi A, Ikeda K, Irieda H, Yoshino K, Yoshida K, Matsumura H, et al. Towards establishment of a rice stress response interactome. PLoS Genet. 2011;7:e1002020.

Sapranauskas R, Gasiunas G, Fremaux C, Barrangou R, Horvath P, Siksnys V. The *Streptococcus thermophilus* CRISPR/Cas system provides immunity in *Escherichia coli*. Nucleic Acids Res. 2011;39(21):9275–82.

Schilperoort RA, Veldstra H, Warnaar SO, Mulder G, Cohen JA. Formation of complexes between DNA isolated from tobacco crown gall tumours and RNA complementary to *Agrobacterium tumefaciens* DNA. Biochim Biophys Acta. 1967;145:523–5.

Schornack S, Meyer A, Römer P, Jordan T, Lahaye T. Gene-for-gene-mediated recognition of nuclear-targeted AvrBs3-like bacterial effector proteins. J Plant Physiol. 2006;163:256–72.

Seo YS, Chern M, Bartley LE, Han M, Jung KH, Lee I, Walia H, Richter T, Xu X, Cao P, Bai W, Ramanan R, Amonpant F, Arul L, Canlas PE, Ruan R, Park CJ, Chen X, Hwang S, Jeon JS, Ronald PC. Towards establishment of a rice stress response interactome. PLoS Genet. 2011;7:e1002020.

Serba DD, Yadav RS. Genomic tools in pearl millet breeding for drought tolerance: status and prospects. Front Plant Sci. 2016;22:1724.

Shan Q, Wang Y, Li J, Zhang Y, Chen K, Liang Z, Zhang K, Liu J, Xi JJ, Qiu JL, et al. Targeted genome modification of crop plants using a CRISPR-Cas system. Nat Biotechnol. 2013;31:686–8.

Shukla VK, Doyon Y, Miller JC, DeKelver RC, Moehle EA, Worden SE, Mitchell JC, Arnold NL, Gopalan S, Meng X, Choi VM. Precise genome modification in the crop species Zea mays using zinc-finger nucleases. Nature. 2009 May;459(7245):437–41.

Singh R, Dangol S, Chen Y, Choi J, Cho YS, Lee JE, Choi MO, Nam-Soo J. *Magnaporthe oryzae* effector AVR-Pii helps to establish compatibility by inhibition of the rice NADP-Malic enzyme resulting in disruption of oxidative burst and host innate immunity. Mol Cells. 2016;39(5):426–38.

Snyter LA, Brooks JE. The characterization and cloning of the Not1 restriction-modification system. Heredity. 1988;61:308.

Sorek R, Kunin V, Hugenholtz P. CRISPR—a widespread system that provides acquired resistance against phages in bacteria and archaea. Nat Rev Microbiol. 2008;6:181.

Srinivasachary HS, Shivayogi S, Vaishali MG, Shashidhar HE, Kumar GK. Genetic analysis of rice blast fungus of southern Karnataka using DNA markers and reaction of popular rice genotypes. Curr Sci. 2002;82:732–5.

Suh JP, Roh JH, Cho YC, Han SS, Kim YG, Jena KK. The Pi40 gene for durable resistance to rice blast and molecular analysis of Pi40-advanced backcross breeding lines. Phytopathology. 2009;99:243–50.

Takan JP, Chipili J, Muthumeenakshi S, Talbot NJ, Manyasa EO, Bandyopadhyay R, Sere Y, Nutsugah SK, Talhinhas P, Hossain M, Brown AE, Sreenivasaprasad S. *Magnaporthe oryzae* populations adapted to finger millet and rice exhibit distinctive patterns of genetic diversity, sexuality and host interaction. Mol Biotechnol. 2012;50:145–58.

Tanaka S, Brefort T, Neidig N, Djamei A, Kahnt J, Vermerris W, Koenig S, Feussner K, Feussner I, Kahmann R. A secreted *Ustilago maydis* effector promotes virulence by targeting anthocyanin biosynthesis in maize. Elife. 2014;3:e01355.

Thomas HR, Percival SM, Yoder BK, Parant JM. High-throughput genome editing and phenotyping facilitated by high resolution melting curve analysis. PLoS One. 2014;9:e114632.

Townsend JA, Wright DA, Winfrey RJ, Fu F, Maeder ML, Joung JK, Voytas DF. High-frequency modification of plant genes using engineered zinc-finger nucleases. Nature. 2009 May;459(7245):442–5.

Van der Hoorn RA, Kamoun S. From Guard to Decoy: a new model for perception of plant pathogen effectors. Plant Cell. 2008;20:2009–17.

Van Larebeke N, Engler G, Holsters M, van den Elsacker S, Zaenen I, Schilperoort RA, Schell J. Large plasmid in *Agrobacterium tumefaciens* essential for crown-gall-inducing ability. Nature. 1974;252:169–70.

Vouillot L, Thélie A, Pollet N. Comparison of T7E1 and surveyor mismatch cleavage assays to detect mutations triggered by engineered nucleases. G3 (Bethesda). 2015;5(3):407–15.

Waltz E. Nonbrowning GM apple cleared for market. Nat Biotechnol. 2015 Apr 1;33(4):326–8.

Wang Y, Cheng X, Shan Q, Zhang Y, Liu J, Gao C, Qiu JL. Simultaneous editing of three homo alleles in hexaploid bread wheat confers heritable resistance to powdery mildew. Nat Biotechnol. 2014;32:947–51.

Wang FJ, Wang CL, Liu PQ, Lei CL, Hao W, Gao Y, Liu YG, Zhao KJ. Enhanced rice blast resistance by CRISPR/Cas9-targeted mutagenesis of the ERF transcription factor gene OsERF922. PLoS One. 2016;11(4):e0154027.

Wawra S, Djamei A, Albert I, Nürnberger T, Kahmann R, van West P. *In vitro* translocation experiments with RxLR-reporter fusion proteins of Avr1b from *Phytophthora sojae* and AVR3a from *Phytophthora infestans* fail to demonstrate specific autonomous uptake in plant and animal cells. Mol Plant-Microbe Interact. 2013;26:528–36.

White FF, Potnis N, Jones JB, Koebnik R. The type III effectors of *Xanthomonas*. Mol Plant Pathol. 2009;10:749–76.

Wirthmueller L, Zhang Y, Jones JD, Parker JE. Nuclear accumulation of the *Arabidopsis* immune receptor RPS4 is necessary for triggering EDS1-dependent defense. Curr Biol. 2007;17:2023–9.

Wright DA, Townsend JA, Winfrey RJ Jr, Irwin PA, Rajagopal J, Lonosky PM, Hall BD, Jondle MD, Voytas DF. High-frequency homologous recombination in plants mediated by zinc-finger nucleases. Plant J. 2005 Nov;44(4):693–705.

Yanik M, Alzubi J, Lahaye T, Cathomen T, Pingoud A, Wende W. TALE-PvuII fusion proteins–novel tools for gene targeting. PLoS One. 2013;8(12):e82539.

Zaenen I, Van Larebeke N, Teuchy H, VanMontagu M, Schell J. Supercoiled circular DNA in crown-gall inducing *Agrobacterium* strains. J Mol Biol. 1974;86:109–27.

Zhang F, Maeder ML, Unger-Wallace E, Hoshaw JP, Reyon D, Christian M, Li X, Pierick CJ, Dobbs D, Peterson T, Joung JK. High frequency targeted mutagenesis in Arabidopsis thaliana using zinc finger nucleases. Proc Natl Acad Sci. 2010 Jun 29;107(26):12028–33.

Zhang S, Wang L, Wu W, He L, Yang X, Pan Q. Function and evolution of *Magnaporthe oryzae* avirulence gene AvrPib responding to the rice blast resistance gene Pib. Sci Rep. 2015;25:11642.

Zhu X, Xu Y, Yu S, Lu L, Ding M, Cheng J, Meng S. An efficient genotyping method for genome-modified animals and human cells generated with CRISPR/Cas9 system. Sci Rep. 2014;4(1):1–8.

Chapter 14
Current Status and Management of Foxtail Millet [*Setaria italica* (L.) Beauv.] Blast Disease

Rajesha G, I. K. Das, Sakthivel Krishnan, and Vilas A. Tonapi

Contents

14.1 Introduction

Foxtail millet [*Setaria italica* (L.) Beauv.] [(synonym: *Panicum italicum* L.)] is one among the small millets commonly called as Italian millet, German millet, and Hungarian millet. It is an important ancient crop of dryland agriculture, grown since 10,500 years ago in China (Yang et al. 2012). This crop is mainly cultivated for food purposes in Asia, whereas in the United States and Europe as fodder for animal feed (Seetharam et al. 1989). Foxtail millet ranks the second position in the total world production among the millets and is an essential staple food crop for millions of

R. G (✉) · I. K. Das · V. A. Tonapi
ICAR-Indian Institute of Millets Research, Rajendranagar, Hyderabad, Telangana, India
e-mail: rajesha@millets.res.in

S. Krishnan
ICAR-Indian Institute of Oilseed Research, Rajendranagar, Hyderabad, Telangana, India

S. C. Nayaka et al. (eds.), *Blast Disease of Cereal Crops*, Fungal Biology,
https://doi.org/10.1007/978-3-030-60585-8_14

people residing in southern Europe and Asia (Marathee 1993) and extensively cultivated in the developing countries in semiarid and arid regions of Africa, Americas, and Asia (Lata et al. 2013). In India, the crop is cultivated in an area of 0.98 lakh ha with the production of 0.56 lakh tons of grain and average productivity of 565 Kg/ha (Hariprasanna 2017). In India it is mostly cultivated in Karnataka, parts of coastal Andhra Pradesh, Uttarakhand, Tamil Nadu, and some pockets of northeastern states (Sharma et al. 2014). Generally, this millet is cultivated all over the world at present, and is most important in China, India, Indonesia, Korean, south-eastern Europe, the United States, and Australia (Sheahan 2014; Taylor 2018).

This crop generally requires short duration, and is highly resilient to drought, physiologically very efficient, and reliable for harvest. In some parts, it is grown as a catch crop when the main crop is failed to utilize the remaining season. The grains are highly nutritious and even well superior to rice and wheat with respect to certain constituents (Upadhyaya et al. 2011) because of its lushness in dietary fiber content, antioxidants, phytochemicals, and polyphenols that are beneficial to human health (Muthamilarasan et al. 2016). The grains contain a greater amount of protein, minerals (calcium, iron, potassium, magnesium, and zinc), and vitamins (Rai 2002). Seetharam et al. (1983) reported 4.0–7.3% of seed oil content in various foxtail millet germplasms. The protein present in foxtail millet is also used as a food component to fight type 2 diabetes and heart diseases (Choi et al. 2005). Hermuth et al. (2016) reported that essential amino acids (threonine, valine, methionine, isoleucine, leucine, and phenylalanine) present in grains of foxtail millet are 65%, 51.1%, and 41% higher as compared to wheat, corn, and rice, respectively.

Climate change is one of the undesirable phases which may intensely affect agricultural production in the coming days. The most serious threats that occur due to climate change are biotic and abiotic stress. Among the biotic stress, diseases play a risk factor for its production, *viz.* blast (*Pyricularia setariae*), rust (*Uromyces setariae-italica*), smut (*Ustilago crameri*), brown spot (*Drechslera setariae*), downy mildew (*Sclerospora graminicola*), udbatta (*Ephelis* sp.), and bacterial leaf blight (*Pseudomonas avenae*) that have been reported in foxtail millet (Das 2017). Among the diseases, the blast caused by *Pyricularia setariae* Nishikado is of moderate importance in India but it may become a major constraint for the production of foxtail millet especially in northern China and India (Nakayama et al. 2005) distressing the production of both forage and grains. During the favorable environmental condition, the disease occurs in severe form and may cause grain loss of up to 60% (Nagaraja et al. 2007; Karthikeyan and Gnanamanickam 2008).

14.2 Distribution

Initially, the blast pathogen was reported by Kawakami (1901–1902) as *P. oryzae* Cav. But this blast pathogen was first reported by Nishikado from Japan in 1917 and identified as *Pyricularia setariae*. In India, this pathogen was reported by McRae (1922) for the first time from Tamil Nadu. Again, Thomas (1940) reported that *P. oryzae*, the cause of paddy blast, could infect *S. italica*, but *Pyricularia* species from *S. italic* is unable to infect rice. The fungus is seed-borne and to some extent soilborne (Palaniswamy et al. 1970). The spread of this pathogen occurs due to the accomplishment of the pathogen in seeds. Later, several countries have reported the occurrence of disease in foxtail millet. In Iran the first report of blast disease was reported by Adel Pordel in 2016 (Pordel et al. 2018).

14.3 Symptoms

The symptoms of blast disease appear on different parts of the plant, viz. leaf, sheath, node, neck, stem, and head. Lower leaves are the most severely affected parts (Gaikwad and D'Souza 1986). The seedling blast is also observed at the seedling stage (Sharma et al. 2014). Symptoms of the disease appear as circular spots with straw-colored centers on leaf blades (Das et al. 2016). The spots are small and scattered, 2–5 mm in diameter, and surrounded by a dark brown margin. The spots coalesce and make the leaves to dry up. When the disease appears in severe form during humid weather conditions, especially with a dense plant stand, the leaves wither and dry as shown in Fig. 14.1.

14.4 The Causal Organism and Host Range

The mycelia of *P. setariae* is thin, hyaline, and straight in young culture whereas as culture becomes old, the mycelium becomes thicker, attains slightly brownish tinge, and swells (Ramakrishnan 1948). The conidiophores emerge through epidermal cells or stomata. Several conidia are formed one after another from the apex of each conidiophore. The production of conidia occurs during high relative humidity and it is released under wind pressure. They are subhyaline, three-celled pyriform, and measure 19–30 μ × 9–15 μ (Kulkarni 1969). Germ tubes are formed from the end cells on germination. Thick-walled, brown, globose chlamydospores are developed at the tips of the germ tubes. The fungus grows well on agar media and host leaf extract.

Fig. 14.1 Foxtail millet sowing leaf blast symptoms

The Pyricularia species has a wild host range belonging to the graminaceae family. These species of Pyricularia were categorized into different species by several authors based on the comparative studies on their pathogenicity factors, host range, mating ability, and isozyme patterns of the extracellular enzyme of pathogens (Kato and Yamaguchi 1980; Matsuyama et al. 1977). In addition to these traditional characteristics, molecular markers such as restriction fragment length polymorphisms (RFLPs) of ribosomal RNA gene (rDNA) and mitochondrial DNA (mtDNA), DNA fingerprints with repetitive DNA, and random amplified polymorphic DNA have been applied to elucidate genetic similarities of *Pyricularia* isolates (Ko et al. 1993; Borromeo et al. 1993; Huff et al. 1994; Valent and Chumley 1991). These studies proposed that *Pyricularia* isolates could be classified into several host-specific groups.

The occurrence and infectivity of this pathogen were not reported in any other crop species as cross infectivity was reported in other species. Many researchers reported that *P. oryzae*, the cause of paddy blast, could infect *S. italica*, but *Pyricularia* species from *S. italica* failed to infect rice crop (Thomas 1940; Ramakrishnan 1948; Wallace 1950). It shows that the pathogen infecting foxtail millet is genetically different from the species infecting other crops. *Pyricularia setariae* pathogen is highly host specific in nature. Showed that *Pyricularia* from *S. italica* resembled *P. setariae* rather than *P. oryzae*. Since *P. setariae* is not associated with the rice, much work on *Pyricularia* from *S. italica* was not done, as compared to *P. oryzae*.

Nishikado (1917) compared host ranges and morphological characters of 16 *Pyricularia* isolates from six host species, and proposed the scientific name *P. setariae* for the three Setaria isolates, mainly because of their specific pathogenicity to foxtail millet. Kato et al. (2000) classified 85 isolates of pathogens into different pathotypes based on the RFLP analysis. He recognized the three isolates from green foxtail that were included and were only pathogenic to foxtail millet among their differential plants. Therefore, pathogenic isolates of foxtail could be classified as in separate pathotypes and named as Setaria pathotypes. Viswanath and Seetharam 1989) reported that *P. setariae* was also found to infect finger millet, pearl millet, and wheat. Furhter Doust, and Kellogg et al. (2002) and Yamagashira et al. (2008) reported that isolates of foxtail millet also infect bristle grass but distinct from *P. grisea*. Whereas, Sharma et al. (2014) reported that *M. grisea* strains from rice or any other hosts do not infect foxtail millet and vice versa.

14.5 Survival and Spread

The pathogen mainly survives in the previous crop residue leftover in the main field and on other cereals including weeds present on the bunds as well as in and around the field. The primary source of inoculum comes from weeds or collateral hosts which belong to Poaceae; when the sporulation occurs on these hosts, the conidia are liberated and spread by wind.

14.6 Disease Cycle and Epidemiology

The blast of foxtail millet is polycyclic in nature and infects all parts of the plant. This fungus produces both sexual spores (ascospores) and asexual spores (conidia). In asexual reproduction, airborne pyriform conidia liberate from conidiophore, land on the surface of the leaves of the plant, and adhere to the surface by producing sticky mucilaginous substance. Later the spore germinates when it gets sufficient moisture especially in the presence of dewdrops, leading to the formation of a germ tube. The germ tube becomes swallow and forms the appressorium which enters the plant through natural openings. The fungal hyphae penetrate into the plant tissue, grow, and eventually produce lesions. The blast lesions become apparent between 72 and 96 h after infection (Agrios 2005). Under higher relative humidity, sporulation of pathogen is high and liberated. Again these conidia start infecting the other plants to continue the next disease cycle. Like this, asexual cycle can be repeated many times during each growing season based on the availability of suitable environmental condition. In the case of sexual reproduction, the production of sexual spores is found inside the specialized fruiting body called perithecium. The asci containing ascospores released from perithecium and hyphal formation after ascospore germination lead to the production of airborne conidia (Fig. 14.2).

14.6.1 Epidemiological Requirements

Many factors influence the development of blast epidemics; mainly it depends on the susceptibility of a variety, availability of primary inoculum load to initiate the disease development, excessive application of nitrogen fertilizer, cloudy and drizzling weather or dew resulting in continuous leaf wetness for more than 10 h, night temperature between 15 and 24 °C, and relative humidity above 90%.

14.7 Integrated Disease Management Approaches

Adaptation of integrated approaches is an effective means to combat the disease. In the practical and majority of cropping systems today, the combination of suitable management practices is the holistic strategy for effective and sustainable management of disease instead of a single method.

14.7.1 Host Plant Resistance

Among the methods of disease management, host plant resistance is one of the simplest, practical, effective, and economical for plant disease management. The use of resistant varieties can not only ensure protection against diseases but also save the

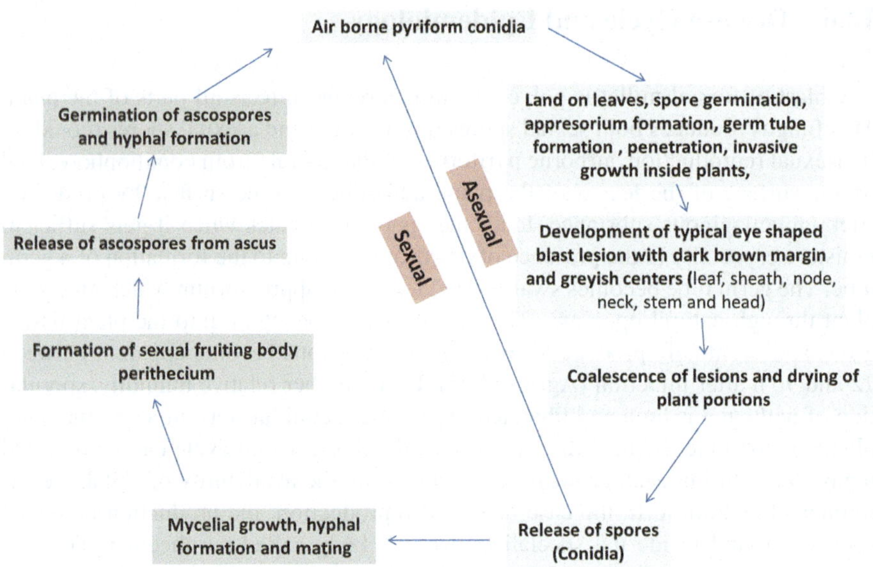

Fig. 14.2 Life cycle of blast fungus *Pyricularia* spp.

time, energy, and money spent on other measures of control. It is very much practical in such cases where chemical control is very expensive and impractical. Breeding for improved blast-resistant varieties is an important goal of foxtail millet improvement programs. Singh et al. (1976) found the varieties SR 118, SR 102, ISc 709, 701, 703, 710, 201, JNSc 33, 56, RS 179, and ST 5307 as resistant to blast in India. Later Sharma et al. (2014) reported that multiple pathotype-resistant accessions identified in the core collection could be used in breeding programs. Among blast-resistant accessions in his study, ISe 376 was found resistant to three of the four foxtail blast isolates tested out of 155 core collections evaluated. Adaptation of integrated approaches is an effective means to combat the disease. Varietal resistance offers one of the most effective means of controlling this disease. Hence, it is necessary to have information on genotypes resistant against the disease. Regarding this, many field screenings have been conducted and have identified the resistant source against the disease in India. But further utilization of unique blast-resistant source in breeding programs is not yet taken up and this needs to be concentrated in future breeding programs on disease resistance.

14.7.2 Cultural Methods

Cultural control methods are very important as it helps the crops to escape from initial pathogen inoculum and the further infection and spread of disease. It involves mainly good agricultural practices right from sowing of seeds to harvest of crop. The important cultural practices include use of disease-free good-quality seeds, proper date of sowing, removal and destruction of weeds and collateral hosts regularly, selection of resistant cultivars, avoiding the use of excess nitrogenous fertilizer, application of N fertilizers in split doses, and crop rotation, all of these singly or in combination.

14.7.3 Chemical Methods

When initial blast spots are seen immediate spraying with effective fungicides like carbendazim 50 WP @ 1 g/L, edifenphos 50 EC @ 1 mL/L, or a combination product of carbendazim + mancozeb @ 1 g/L of water has to be resorted to intercept further development of the disease (Konda et al. 2016). Chemical methods for disease management are followed in high-disease-pressure-prone areas. Seed treatment with captan or carbendazim or thiram or tricyclazole at 2.0 g/kg seed can be used to reduce initial disease incidence at nurseries. During low infection at tillering stage, systemic fungicides such as pyroquilon and tricyclazole are possible chemicals for controlling the disease. Spraying of tricyclazole at 1 g/L of water, edifenphos at 1 mL/L of water, or carbendazim at 1.0 g/L is useful. For successful control 2–3 sprays of chemicals may be used from seedling to booting stage of crop.

14.7.4 Biological Control

Biological control of plant pathogens using antagonistic and beneficial microbes has been found promising in recent days. Many fungal and bacterial diseases have been successfully controlled through biocontrol agents worldwide. Many of the beneficial bacterial genera like *Bacillus* spp. (Chen et al. 2019) and *Pseudomonas* spp. (Sakthivel and Gnanamanickam 1987; Gnanamanickam 2009), and also few yeast species, like *Streptomyces* spp. (Law et al. 2017), were found effective in the management of rice and finger millet blast diseases caused by *Pyricularia oryzae* and *Pyricularia grisea*, respectively. In case of foxtail millet blast, few studies showed *in vitro* potential of *Trichoderma* spp. and *Bacillus* spp. against *P. setariae*; however, not many studies have been reported on the field performance. Hence further studies are needed to strengthen the recommendations on biological control of blast disease in foxtail millet.

References

Agrios G. Plant Pathology. 5th edition, Academic Press. 2005 p. 952.

Borromeo ES, Nelson RJ, Bonman JM, Leung H. Genetic differentiation among isolates of *Pyricularia* infecting rice and weed hosts. Phytopathology. 1993;83:393–9.

Chen WC, Chiou TY, Delgado AL, Liao CS. The control of rice blast disease by the novel biofungicide formulations. Sustainability. 2019;11(12):3449.

Choi Y, Osada K, Ito Y, Nagasawa T, Choi M, Nishizawa N. Effects of dietary protein of Korean foxtail millet on plasma adiponectin, HDL-cholesterol and insulin levels in genetically type 2 diabetic mice. Biosci Biotechnol Biochem. 2005;69:31–7.

Das IK. Millet diseases: current status and their management. 2017. p. 291–322. In: Patil JV, editor. Millets and Sorghum biology and genetic improvement. Wiley Blackwell. p. 463.

Das IK, Nagaraja A, Tonapi VA. Diseases of millets—a ready reckoner. Indian Institute of Millets Research, Rajendranagar, Hyderabad 500030, Telangana. 2016. 67 p. ISBN: 81-89-335-59-6.

Doust AN, Kellogg EA. Inflorescence diversification in the panicoid "bristle grass" clade (Paniceae, Poaceae): evidence from molecular phylogenies and developmental morphology. Am J Bot. 2002;89:1203–1222.

Gaikwad AP, D'Souza TF. Studies on blast disease of Rala *Setaria italica*. J Maharashtra Agric Univ. 1986;11:108–9.

Gnanamanickam SS. Biological control of rice blast. Dordrecht: Springer; 2009. p. 53–65.

Hariprasanna K. Foxtail millet, *Setaria italica* (L.) P. Beauv. 2017. p. 112–49. In: Patil JV, editor. Millets and Sorghum biology and genetic improvement. Wiley Blackwell. p. 463.

Hermuth J, Janovska D, Cepkova PH, Ustak S, Strasil Z, Dvorakova Z. Sorghum and foxtail millet—promising crops for the changing climate in central Europe. In: Konvalina P, editor. Alternative crops and cropping systems. Rijeka: In Tech; 2016. p. 3–28. https://doi.org/10.5772/62642.

Huff DR, Bunting TE, Plumley KA. Use of random amplified polymorphic DNA markers for the detection of genetic variation in *Magnaporthe poae*. Phytopathology. 1994;84:1312–6.

Karthikeyan V, Gnanamanickam S. Biological control of Setaria blast (*Magnaporthe grisea*) with bacterial strains. Crop Prot. 2008;27(2):263–7.

Kato H, Yamaguchi T. Host range and interrelations of Pyricularia species from various cereals and grasses. Proc Kanto-Tosan Plant Prot Soc. 1980;27:14–5.

Kato H, Yamamoto M, Yamaguchi-Ozaki T, Kadouchi H, Iwamoto Y, Nakayashiki H, Tosa Y, Mayama S, Mori N. Pathogenicity, mating ability and DNA restriction fragment length polymorphisms of Pyricularia populations isolated from Gramineae, Bambusideae and Zingiberaceae plants. J Gen Plant Pathol. 2000;66:30–47.

Ko JH, Hwang BK, Hwang BG. Restriction fragment length polymorphisms of mitochondrial and nuclear DNAs among Korean races of *Magnaporthe grisea*. J Phytopathol. 1993;138:41–54.

Konda S, Nagaraja A, Nagamma G, Sangeetha PS, Patil S, Dev D, Anjum SS. *In-vitro* evaluation of bio-agents and fungicides against leaf blast (*Pyricularia setariae*) in foxtail millet [*Setaria italica* (L.) Beauv.]. J Pure Appl Microbiol. 2016;10(1):489–96.

Kulkarni NB. Comparative studies of four isolates of *Pyricularia setariae* Nishikado from setaria species in India. Mycopathol Mycol Appl. 1969;38:299–303. https://doi.org/10.1007/BF02052682.

Lata C, Gupta S, Prasad M. Foxtail millet: a model crop for genetic and genomic studies in bioenergy grasses. Crit Rev Biotechnol. 2013;33:328–43. https://doi.org/10.3109/07388551.2012.716809.

Law JWF, Ser HL, Khan TM, Chuah LH, Pusparajah P, Chan KG, Goh BH, Lee LH. The potential of Streptomyces as biocontrol agents against the rice blast fungus, *Magnaporthe oryzae* (*Pyricularia oryzae*). Front Microbiol. 2017;8:3.

Marathee JP. Structure and characteristics of the world millet economy. In: Riley KW, Gupta SC, Seetharam A, Mushonga JN, editors. Advances in small millets. New Delhi: Oxford & IBH; 1993. p. 159–78.

Matsuyama N, Kato H, Yamaguchi T. Comparison of the isozyme patterns of extracellular enzymes in *Pyricularia* strains from gramineous and zingiberaceous plants. Ann Phytopathol Soc Jpn. 1977;43:419–25.

McRae W. Report of the imperial mycologist. Sci Rep Agric Res Inst, Pusa (1921–22). 1922. p. 44–50.

Muthamilarasan M, Dhaka A, Yadav R, Prasad M. Exploration of millet models for developing nutrient rich graminaceous crops. Plant Sci. 2016;242:89–97.

Nagaraja A, Kumar J, Jain AK, Narasimhzdu Y, Raghuchander T, Kumar B, Gowda BH. Compendium of small millets diseases. Project Coordinator Cell, All India Coordinated Small Millets Improvement Project, UAS, GKVK Campus, Bengaluru. 2007. p. 80.

Nakayama H, Nagamine T, Hayashi N. Genetic variation of blast resistance in foxtail millet (Setaria italica (L.) P. Beauv.) and its geographic distribution. Genetic Resources and Crop Evolution. 2005;52(7):863–868.

Nishikado Y. Studies on rice blast fungus. Iohara Inst. land hw. Forsch. Ber. 1917;1: 179– 219.

Palaniswami A, Govindaswami CV, Vidyasekaran P. Studies on blast diseae of tenai. Madras Agril. J. 1970;57: 686–693.

Pordel A, Tharreau D, Cros-Arteil S, Shams E, Moumeni A, MirzadiGohari A, Javan-Nikkhah M. *Pyricularia oryzae* causing blast on foxtail millet in Iran. Plant Dis. 2018;102(9):1853. https://doi.org/10.1094/PDIS-01-18-0091-PDN.

Rai M. Nutritive cereals. In: Survey of Indian agriculture. Chennai: The Hindu; 2002. p. 59–62.

Ramakrishnan. Studies on the morphology, physiology and parasitism of the genus Pyricularia in Madras. Proc Indian Acad Sci Sect B. 1948;27(6):174–93.

Sakthivel N, Gnanamanickam SS. Evaluation of Pseudomonas fluorescens strains for suppression of sheath-rot disease and enhancement of grain yields in rice, *Oryza sativa*, L. Applied and Environmental Microbiology. 1987;3:2056–2059.

Seetharam A, Aradhya KM, Laxminarayana MR. Variation for seed oil content in a world collection of foxtail millet germplasm. SABRAO J. 1983;15:99.

Seetharam A, Riley KW, Harinarayana G. Small millets in global agriculture. New Delhi: Oxford & IBH Publishing Co.; 1989.

Sharma R, Girish AG, Upadhyaya HD, Humayun P, Babu TK, Rao VP, Thakur RP. Identification of blast resistance in a core collection of foxtail millet germplasm. Plant Dis. 2014;98:519–24.

Sheahan CM. Plant guide for foxtail millet (*Setaria italica*). Cape May: USDA-Natural Resources Conservation Service; 2014.

Singh BP, Pad BS, Nema AG, Deshkar MV. Varietal reaction to blast of Kangni. Curr Res. 1976;1:13.

Taylor JRN. Sorghum and millets pages 1–21. In: Taylor JRN, Duodu KG. 2nd ed. Sorghum and millets: chemistry, technology and nutritional attributes. Elsevier Inc. in Cooperation with AACC International; 2018. p. 453. https://doi.org/10.1016/b978-0-12-811527-5.00001-0.

Thomas KM. Detailed administration report of the Government Mycologist for the year 1940 Rep. Dep Agric Madras. 1940;41:53–74.

Upadhyaya HD, Ravishankar CR, Narasimhudu Y, Sharma NDRK, Singh SK, Varshney SK, Reddy VG, Singh S, Parzies HK, Dwivedi SL, Nadaf HL, Sahrawat KL, Gowda CLL. Identification of trait specific germplasm and developing a mini core collection for efficient use of foxtail millet genetic resources in crop improvement. Field Crops Res. 2011;124:459–67.

Valent B, Chumley FG. Molecular genetic analysis of the rice blast fungus, *Magnaporthe grisea*. Annu Rev Phytopathol. 1991;29:443–67.

Viswanath S, Seetharam A. Diseases of small millets and their management in India. In: Seetharam A, Riley KW, Harinarayana G, editors. Small millets in global agriculture. New Delhi: Oxford & IBH Publishing Co. Pvt. Ltd; 1989. p. 237–53.

Wallace GB. Annual report of the Plant Pathologist 1948-Rep. Dep Agric Tanganyika 1948. 1950. p. 145–7.

Yamagashira A, Iwai C, Misaka M, Hirata K, Fujita Y, Tosa Y, Kusaba M. Taxonomic characterization of Pyricularia isolates from green foxtail and giant foxtail, wild foxtails in Japan. J Gen Plant Pathol. 2008;74:230–241.

Yang X, Wan Z, Perry L, Lu H, Wang Q, Zhao C. Early millet use in Northern China. Proc Natl Acad Sci U S A. 2012;109(10):3726–30.

Correction to: Blast Disease of Cereal Crops: Evolution and Adaptation in Context of Climate Change

S. Chandra Nayaka, Rajashekara Hosahatti, Ganesan Prakash, C. Tara Satyavathi, and Rajan Sharma

Correction to:
S. C. Nayaka et al. (eds.), *Blast Disease of Cereal Crops*, Fungal Biology, https://doi.org/10.1007/978-3-030-60585-8

This book was inadvertently published with incorrect co-editor name as "Ganesh" instead of "Ganesan". This is now corrected in all the renditions.

The updated version of the book can be found at
https://doi.org/10.1007/978-3-030-60585-8

S. C. Nayaka et al. (eds.), *Blast Disease of Cereal Crops*, Fungal Biology,
https://doi.org/10.1007/978-3-030-60585-8_15

Index

S. C. Nayaka et al. (eds.), *Blast Disease of Cereal Crops*, Fungal Biology,
https://doi.org/10.1007/978-3-030-60585-8

The manufacturer's authorised representative in the EU is Springer Nature Customer Service Centre GmbH, Europaplatz 3, 69115 Heidelberg, Germany. If you have any concerns regarding our products, please contact ProductSafety@springernature.com

Printed and bound by CPI Group (UK) Ltd, Croydon, CR0 4YY
28/09/2026
02224089-0008